FINDING ORDER IN NATURE

JOHNS HOPKINS

INTRODUCTORY STUDIES

IN THE HISTORY

OF SCIENCE

Mott T. Greene

and Sharon Kingsland,

Series Editors

Finding Order in Nature

*The Naturalist Tradition
from Linnaeus to E. O. Wilson*

Paul Lawrence Farber

THE JOHNS HOPKINS UNIVERSITY PRESS

BALTIMORE AND LONDON

© 2000 The Johns Hopkins University Press
All rights reserved. Published 2000
Printed in the United States of America on acid-free paper
9 8 7 6 5 4 3 2 1

The Johns Hopkins University Press
2715 North Charles Street
Baltimore, Maryland 21218-4363
www.press.jhu.edu

Library of Congress Cataloging-in-Publication Data will be found at the end of
this book.
A catalog record for this book is available from the British Library.

ISBN 0-8018-6389-9
ISBN 0-8018-6390-2 (pbk.)

To Fritz Marti

Contents

Acknowledgments

This project would not have gotten off the ground had it not been for two individuals. In response to a remark of mine about how long it would be before I would be "ready" to embark on such a study, Fred Churchill (who had pointed me long ago in the direction of the history of natural history) said, "Just do it." Mott Greene, not long after the discussion with Fred, asked me if I would be interested in contributing a volume to a new series that the Johns Hopkins University Press was planning. So, I decided the time had come. Robert J. Brugger, at the Johns Hopkins University Press, was receptive and encouraged me throughout the project.

Natural history has long fascinated me, and most of my professional career has been spent studying various aspects of its development. Many people have helped me over the years, and it would be impolitic to try to reduce the list of names to a convenient size for inclusion. Some people, however, have been so central in my intellectual development for so many years that it would be even more impolitic not to mention them. Phil Sloan, Maggie Osler, Jim Morris, Brookes Spencer, Bart Hacker, Gar Allen, David Allen, Keith Benson, Jane Maienschein, Chip Burckhardt, Joel Hagen, Marsha Richmond, and Mike Osborne have spent hours discussing the history of science with me and have greatly contributed to what I know and how I see the subject. Tom Franzel gave an early version of this book a careful reading and made many useful suggestions. Sharon Kingsland read a later version and generously took the time to trash a lot of it, thus saving me and the reader a lot of anguish. Bob Brugger picked up where Sharon left off, also to the reader's benefit. The office staff in the Department of History at Oregon State University provided valuable assistance—Ginny Domka, Sharon Johnson, and Marilyn Bethman efficiently took care of innumerable tasks connected with the project. My colleagues in the Department of History have provided a stimulating and supportive environment in which to work.

I have conducted most of my research in libraries, and I would like to acknowledge the professional, friendly, and helpful library staffs at Oregon State

University, the University of Washington, Harvard University, the University of California at Berkeley, the Linnean Society, the British Library, the British Museum (Natural History), the Muséum national d'histoire naturelle, and the Bibliothèque nationale.

My parents provided me with opportunities, encouragement, and support early in my career, and everything I have done has built on that. My intellectual parents, Fred Churchill, Sam Westfall, and Joe Schiller, worked hard at trying to educate me, and I continue to marvel at the time and effort they so freely gave. My intellectual partner and companion, Vreneli Farber, provides me with an emotional center and has acted as an invaluable, critical sounding board. My children, Benjamin and Channah, give me hope for the future.

This book is dedicated to the memory of my father-in-law, Fritz Marti, an inspiring intellectual whom I had the good fortune to know for many years.

FINDING ORDER IN NATURE

Introduction

In the second chapter of *Genesis* we read that "out of the ground the LORD God formed every beast of the field and every bird of the air, and brought them to the man [Adam] to see what he would call them; and whatever the man called every living creature, that was its name" (Gen. 2.19 RSV). Jewish and Christian theologians have traditionally interpreted this oft-quoted passage to mean that God thereby gave man (still without a wife, but shortly to have one) dominion over nature. Humans thereafter legitimately possessed the right to control the natural world and to exploit it for their own use. Those concerned with the origins of the current environmental degradation view that dominion with considerable ambivalence, and some cite this alleged transfer of authority as a root cause of the attitudes responsible for our long history of mismanagement of global resources.

Less controversial, but equally significant, the Genesis story also reflects the long-standing importance of naming and characterizing things found in nature. That one of Adam's first tasks consisted of naming animals should not surprise us. We all possess a curiosity about the natural objects on Earth—animals, plants, minerals—and for good reason; they are a source of food, medicine, clothing, shelter, and entertainment. Anthropologists, who study different cultures of the world, find that all peoples name and categorize the objects in their environment. And most, if not all, cultures share commonsense ways of conceptualizing the natural world. The simple notion that living things that fly constitute a natural group, for example, is shared by second-graders in Colorado and village elders in southeast Asia.

Naming and categorizing has concerned humans since ancient times, as the Hebrew texts attest. Whether for the most basic requirements of communal life or for the most sophisticated scientific exchanges, we have wanted to communicate information that we have gained about the world. Starting in the eighteenth century, however, a particular approach to this activity emerged as a scientific discipline in Europe and has continued to the present day, the modern tradition of *natural history.* What distinguishes natural history from the "folk biology" of earlier studies is the attempt of naturalists to group animals, plants, and minerals according to shared underlying features and to use rational, systematic methods to bring order to the otherwise overwhelming variation found in nature. Although bats are living animals and fly, naturalists do not consider them "birds" because bats share certain characteristics with other mammals. Nor do naturalists consider a simple alphabetical list of animals a viable option for classifying them, given the enormous number of known animals (750,000 insects alone). Instead, naturalists have been working since the 1700s to document the natural world, systematically naming and organizing the myriad forms found there, as they attempted to discern an underlying order. Although individuals before the eighteenth century pursued similar goals, no large-scale, sustained, and organized effort had existed until then.

In the discipline of natural history, researchers systematically study natural objects (animals, plants, minerals)—naming, describing, classifying, and uncovering their overall order. They do this because such work is an essential first step before other, more complex analyses can be undertaken. We cannot start discussing a wetland, or the interactions within it, until we know something about what is there. Nor can we intelligently talk about the effect of an event on a particular environment until we have a sense of the specific kinds of organisms that inhabit it. Natural history does more, however, than just construct catalogs and field guides, important as these are. It explores broader issues: How do all the pieces fit together? What interactions can we discover? What changes? What responsibilities does our knowledge confer upon us?

This book traces the fascinating story of the study of the natural world that began in the eighteenth century and has since captured the interest of an ever-widening circle of enthusiasts. Eighteenth-century society lavished attention on natural history. The second most frequently owned item in private libraries in France at that time was the naturalist Buffon's monumental 36-volume encyclopedia of animals. Cultured gentlemen and ladies normally owned collections of stuffed birds and of shells, with the size of their "cabinet" (as these collections were called) often reflecting their wealth, taste, and level of refinement.

Interest in natural history, however, extended beyond what was merely fashionable. Beginning in the 1700s and extending well into the 1800s, major European powers engaged in a worldwide scramble to identify natural products of economic importance. Prime ministers believed that the fate of empires rested on identifying, cultivating, and transporting specific plants, such as tea shrubs and rubber trees. Thomas Jefferson sent Lewis and Clark across the North American continent in part to survey the economic potential of the natural products in the newly acquired Louisiana Purchase.

While imperial rivalries sent soldiers and explorers to steaming jungles and exotic highlands, a different but equally ferocious contest took place in the western badlands of America. The quest for dramatic fossil skeletons, particularly dinosaurs, captivated a generation of naturalists in the United States. It resulted in a competition for bones in the 1800s that rivaled battles fought among the Fifth Avenue "robber barons" over coal, iron ore, and oil. The public found the Jurassic treasures fascinating. To illustrate the point one need only consider that, by the early twentieth century, more people visited natural history museums to see these prehistoric remains than attended football games.

Natural history also provided the scene for competition among ideas. Conflict between religious and secular views has often fixed on interpretations of nature. Society on both sides of the Atlantic, for example, argued bitterly over the implications of Darwin's theory of evolution. For some the theory threatened to undermine accepted religion; for others it opened up the possibility of revitalizing what they took to be a declining and outmoded set of religious opinions. As in the searing debates over slavery, families literally fragmented because of conflicting positions on evolution. Respectable men hurled insults at one another at nineteenth-century scientific meetings (on one occasion the intensity of the discourse caused a woman to faint).

The story of natural history does not end with the heyday of museum attendance early in this century or with the breakup of colonial empires. Problems in today's world drive current research in natural history. Much of the impetus for earlier natural history arose from European exploration of exotic regions of the globe. Recent development, primarily to stimulate economic growth, has destroyed many of the sites, such as the coastal forests of Brazil, that formerly lured naturalists from the comforts of home and drew them to dangerous expeditions (from which many did not return alive). Naturalists worldwide fear that the pace of development threatens irreparable damage to these formerly pristine locales and to the associated rich diversity of animal and plant life on much of the planet. Exploitation of tropical rain forests destroys about 76,000 square miles per year—roughly the size of the entire country of Costa Rica. Naturalists such as E. O. Wilson, who are deeply concerned about the issue, point out that the problem is compounded by our lack of knowledge. We are wiping out unknown species, and consequently we have no idea of the potential value of what is now gone forever. Scientists, politicians, and economists differ significantly on what actions should be taken to arrest the loss of biodiversity. At international meetings, however, they do agree that, as gargantuan as the task may appear, a first step would be to complete a basic inventory of Earth's species. The environmental, economic, political, and social well-being of humans may depend upon the success of such initiatives.

In spite of natural history's close tie to the pressing ecological and environmental issues of today, science writers and other commentators in this "high-tech" age occasionally treat the subject primarily as a beginning stage in the investigation of the natural world; being a naturalist means merely to name and describe things found in nature. They patronizingly treat natural history as old-fashioned; a pastime that conjures up images of men in knickers carrying butterfly nets or Victorian ladies with plant presses. Research into the history of the discipline however, quickly dispels such a simplistic caricature. To be sure, naming, describing, and classifying continues to be a basic activity that serves as a foundation for the study of nature. The quest for insight into the order of nature, however, leads naturalists beyond classification to the creation of general theories that explain the living world. Those naturalists who focus on the order of nature inquire about the ecological relationships among organisms and also among organisms and their surrounding environments. They ask fundamental questions of evolution, about how change actually occurs over short and long periods of time. Many naturalists are drawn, consequently, to deeper philosophical and ethical issues: What is the extent of our ability to understand nature? And, understanding nature, will we be able to preserve it? Naturalists question the meaning of the order they discover and ponder our moral responsibility for it.

So, does natural history mean mere butterfly and flower collecting? Only in the sense that Alfred Tennyson referred to when he wrote, in "Flower in the crannied wall":

> Flower in the crannied wall
> I pluck you out of crannies
> I hold you here, root and all, in my hand,
> Little flower—but if I could understand
> What you are, root and all, and all in all,
> I should know what God and man is.

In the eighteenth century Buffon and the Swedish botanist Linnaeus, along with a number of other students of nature, established a coherent tradition of natural history. We trace that tradition to see how it expanded and interacted with other traditions in the life sciences, examining some of its major achievements and considering its present state in the world of science. We note the extent to which it reflected the culture of the times and to what degree it had its own history. In exploring these topics we also examine the institutions in which naturalists performed their research and the source of their funding. We see how natural history has yielded the major unifying theory of the life sciences, uncovered some of the deepest insights into nature, led to concern for the environment, and attracted public interest for more than two and a half centuries.

Fascination with nature led some naturalists to relinquish the comforts of home for the hardships and danger of fieldwork; it drove others to spend days, evenings, nights examining data. Indeed, this tradition has inspired, enlightened, and delighted its practitioners and their audience.

1 Collecting, Classifying, and Interpreting Nature

Linnaeus and Buffon, 1735–1788

Natural history emerged in its modern form as a scientific subject in the eighteenth century. Although many people took part in the enterprise, two were central in defining it and giving it direction: the Swedish botanist Carl Linnaeus and a French nobleman and student of nature, Georges Louis Leclerc, comte de Buffon. They came to natural history from different backgrounds and brought to it different perspectives. At the time, no formal training in natural history was available. Universities did not include it as a subject of study, nor did anyone consider it a profession or an occupation.

Linnaeus received some knowledge of natural history by way of the related discipline of medicine. Medical education included the study of anatomy, physiology, and medical botany, and consequently served as a common path to natural history. Many of the early naturalists had similar experiences. Buffon, in contrast, had a general interest in science and what today we would call forestry. Although the two naturalists approached nature from dissimilar perspectives (and harbored professional jealousies), their work came to serve as a foundation for modern natural history. The combined result of their efforts was the development of principles by which to rationally name and classify the natural products of the entire globe. Equally important, Linnaeus and Buffon sought to understand what they believed to be an overarching natural order, bound by specific—and discernible—laws.

Linnaeus

Medical education in Europe in the 1700s reflected its medieval origins. Major countries such as England and France rarely had more than one or two modestly sized medical schools. In these institutions education stressed texts rather than direct experience. Many offered a degree without requiring a person to have studied there, provided that the individual could pass an exam and present an original thesis on a medical topic. Students often had few attractive options, and those from smaller European nations often had to travel to foreign countries to study. Many went to Dutch schools, which were the most celebrated of the time.

During the early eighteenth century, the United Netherlands consisted of seven semi-independent states, six of which had universities. The University of Leiden, the oldest and best-known in the Netherlands, was in the state of Holland and was internationally renowned for its medical training. It had high fees, however, and stiff requirements for a degree. Leiden did not permit students trained elsewhere to obtain a degree by simply presenting a previously prepared thesis and passing an exam. This was in contrast to other medical faculties. For example, at Harderwijk University, in the state of Gelderland, medical candidates could acquire a degree in only a week's time—and at a substantially reduced rate. So, in 1735, Carl Linnaeus, son of a Swedish village clergyman, traveled to Harderwijk University with the goal of obtaining a medical degree. By June 23, six days after his arrival there, he was a Doctor of Medicine.

The twenty-eight-year-old Linnaeus had studied medicine in Lund and Uppsala (although because of the pathetic state of medical education at both Swedish universities he was largely self-taught). He had brought with him to Harderwijk a thesis entitled "A new hypothesis as to the cause of intermittent fevers," which argued that certain fevers resulted from living on clay soils. Linnaeus aspired to a career back in Sweden, and for this he considered a medical degree from a prestigious Dutch university to be critical. Before leaving Sweden he had proposed to the eighteen-year-old daughter of the town physician in the mining center of Falun. The significant dowry his wife would bring to the marriage would help him get established.

Of greater significance to his professional aspirations was that Linnaeus brought to the Netherlands a set of his writings which so greatly impressed an influential circle of Dutch physicians and amateur naturalists that they persuaded him to stay in the Netherlands for three years. During those years he published his earlier writings, along with several newer manuscripts. It was a remarkable period, for in these works he sketched many of the basic ideas he would develop for the rest of his rich and productive life.

Linnaeus concerned himself primarily with the naming and classifying of natural objects. His interest in these activities reflected their importance to the study of natural history in Linnaeus's time: Europeans each year encountered thousands of new species of animals and plants, plus numerous new rocks and minerals. For decades, the botanical gardens in Amsterdam and Leiden had been major centers for receiving plants from Dutch colonial and trading voyages. Many of these exotic plants from Africa, the New World, the Pacific Islands, and Asia were unknown to European science. Naturalists examined these specimens in order to document the Creation and to keep better track of potentially valuable natural products. Along with their French and British counterparts, Dutch merchants and bankers strove to expand their interests around the world. They wisely encouraged the growth of natural history for practical reasons. At the same time, Europeans recorded local species in ever greater detail.

Linnaeus had firsthand experience with the riches of new specimens. Three years before he went to the Netherlands he secured a grant from the Swedish Royal Society of Science to explore the largely unknown natural history of Lapland. For five months he traveled, observed, and collected animals, plants, and minerals in the far North. Later, while in Amsterdam in 1737, he published a botanical account of his trip, the *Flora Lapponica*. On the Lapland expedition he gained direct knowledge of an exotic habitat and a sense of the enormous physical difficulties facing field naturalists. When an influential medical figure asked him to travel to southern Africa to collect plants for Dutch collections (with the added bait of a possible professorship upon his return), Linnaeus turned the offer down. He had a more comfortable alternative, one that would extend his training in natural history. For two years after receiving his medical degree, he served as the superintendent of the garden (and as house physician) to George Clifford, a wealthy financier and director of the Dutch East India Company. The garden and its hothouses contained specimens from southern Europe, Asia, Africa, and the New World. A private zoo housed a dazzling set of exotic animals ranging from tigers to rare birds.

Linnaeus's experiences in Lapland and in Clifford's gardens gave him a vivid sense of the rapidly developing richness of natural history. Though exciting, the new material did raise problems. Foremost, both the exotic and local material presented a confusing picture because much of it did not easily fit into older classification systems. With no standardized procedure for naming plants, animals, and minerals, authors often gave different names to the same plant. They also sometimes failed to recognize male, female, and juvenile forms of the same animal and named them as three different species.

The first manuscript that Linnaeus published after his doctoral thesis consisted of just twelve printed pages. In the *Systema naturae* (1735), he outlined a general system that he believed would bring order to natural history, a task he considered critical. "The first step in wisdom is to know the things themselves," he wrote in his opening remarks. "This notion consists in having a true idea of the objects; objects are distinguished and known by classifying them methodically and giving them appropriate names. Therefore, classification and name-giving will be the foundation of our science."*

The *Systema naturae* proposed a new system of classification for plants, animals, and minerals. The most original and influential section contained a sexual system of classification for plants. Although the ancients had not understood that plants reproduce sexually, European naturalists by the end of the seventeenth century did. Linnaeus created a brilliantly simple hierarchical system that arranged plants into twenty-four classes according to the number and relative position of their stamens (male parts). He broke down the classes into sixty-five orders, primarily on the basis of the number and position of the pistils (female parts). Using other characteristics, Linnaeus went on to distinguish particular genera, consisting of groups of species with similar characteristics, and even more particular species. The system's simplicity and relative ease of application made it appealing. He used the system in his flora of Lapland and in the splendid catalog he published of Clifford's garden (*Hortus Cliffortianus,* 1738).

Compare Linnaeus's system to that of Joseph Pitton de Tournefort, the famous seventeenth-century French botanist. Tournefort believed that anyone who was serious about the subject should be able to memorize the 698 natural genera that encompassed the 10,000 species then known. By contrast, Linnaeus provided amateurs, travelers, and gardeners with a simpler and more practical method. Acknowledging that his method did not reflect any "real" order in nature, Linnaeus believed that naturalists nevertheless should use his "artificial" system until he developed one that actually conveyed God's plan in nature. He worked the rest of his life at constructing such a "natural" system but was, in the end, unable to formulate one satisfactorily. The sexual system, in the meantime, was widely accepted throughout most of Europe.

In his classification, Linnaeus used terminology that reflected his cultural background. Instead of employing terms like *stamen* or *pistil,* he chose the Greek for "husband" (*andria*) and "wife" (*gynia*). The names of the classes, for example, were *monandria, diantria, triandria,* and so on, reflecting the various types of "marriages" in plants. With the exception of the *monandria,* in which there was one husband and one wife, the others involved multiple so-called husbands, concubines, and other decidedly irregular arrangements. Although some naturalists were shocked by Linnaeus's sexual imagery, his terms stuck.

More important than Linnaeus's use of metaphor were his new rules for

*Carolus Linnaeus, *Systema naturae: Facsimile of the First Edition* (1735; Nieuwkoop: B. De-Graaf, 1964), 19.

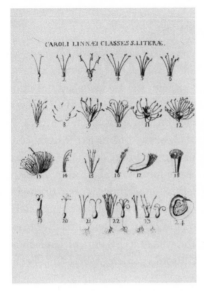

A   B

Naming Plants and Animals

Classification systems fall into two general categories: artificial and natural. An *artificial system* is a means of organizing and retrieving information and makes no claims about the intrinsic or actual relations among groups that the system defines and orders. Descriptive guides to birds or to wild flowers often rely on artificial classifications—on color alone, for example. Linnaeus achieved great fame for his sexual system of classification (*A*). He divided plants into twenty-four classes based on the number of stamens (male sex organ) of the flower or the stamens' positions or relationships (e.g., four long stamens and two short ones). The first eleven classes, for example, are defined by the number of stamens (one, two, etc.).

A *natural system,* on the other hand, attempts to reflect actual relationships in nature. Buffon believed that he had uncovered among quadrupeds a natural order that reflected historical changes they had undergone. He explained the close anatomical similarity of the horse, zebra, and ass by hypothesizing that they were all descendants of an original stock of horses. A comparison (*B*) of the horse (*top*) and the ass (*bottom*) skeleton shows their close similarity.

■   *A,* From G. D. Ehret's plate (1736) appended to Carolus Linnaeus, *Systema naturae* (Leiden: Haak, 1735). *B,* Georges Louis Leclerc, comte de Buffon, *Oeuvres complètes de Buffon* (Brussels: Lejeune, 1828), vol. 6, pl. 10.

nomenclature, or naming plants. Previously the scientific names of plants consisted of two parts: a word (or words) denoting a group of plants, and then a string of characteristics that distinguished the plant from other similar ones. Because no agreed-upon list of names existed and because over the years writers had used different characteristics to name the same plant, considerable confusion had ensued. Linnaeus's reform made plant names more like people's names: a single name common to all the species in a genus, and another, specific name that distinguished the species from others in the genus.

The basic ideas of Linnaeus's binomial (two-word) nomenclature appeared in a manuscript he published in 1736. He later expanded on his principles and used them in his *Species plantarum* (1753), which recorded all known species of plants. The practice quickly caught on. To this day naturalists use the *Species plantarum* (along with the fifth edition of his *Genera plantarum*) as the starting point for botanical nomenclature. Linnaeus also set down rules for selecting names. The names of genera, for example, have Greek or Latin roots only and may not be compounds of two words or commemorate saints or people unconnected with science.

For Linnaeus the naming and ordering of the products of Creation linked the study of nature with the worship of God. Linnaeus's conception of order reflected his vision of Creation as a balanced and harmonious system. Classification, he thought, could reflect that harmony. In his later writings Linnaeus also described a general balance of nature. Every plant and animal fills a particular place in the network of life and helps maintain that network. Carnivores, he observed, daily destroy animals that if unchecked, would reproduce so quickly as to outstrip their sources of food. Such intricate relationships offered proof of a divinely sanctioned balance. The reciprocal relationship of predator and prey linked each in the overall harmonious, static system. Linnaeus believed that the first species had the same relationships in nature as they do now, even after dispersing from their place of creation to their assigned regions, where they have been found ever since. Linnaeus also initially insisted that the species themselves had not changed since their creation, but he later modified this view to accept the idea that hybridization in time had produced new species from the original ones.

Linnaeus stressed the abundance of nature, and he endeavored to catalog it as fully as possible. In the Netherlands he examined magnificent public and private collections, and upon his return to Uppsala he continued his study in the university gardens and created a sizable personal collection. However grand European collections appeared, Linnaeus knew they were nowhere near complete, and so he corresponded with naturalists throughout the world who were eager to have him include their findings in the successive editions of his *Systema naturae*. In his desire to extend his grasp worldwide he actively encouraged his students to undertake extensive voyages to help complete his catalog of life. These explorations afforded adventurous naturalists excitement and challenge. Linnaeus called these students his "apostles." They amassed great collections, and their work extended botanical knowledge. For example, Daniel Solander sailed with the English explorer Captain James Cook on Cook's first voyage around the world. Others traveled to North America, South America, Asia, and throughout the Pacific and returned with impressive natural history collections.

Eighteenth-century travel, exhilarating as it may have been, posed serious risks. Linnaeus had recommended his favorite pupil, Pehr Löfling, to the Spanish ambassador in Stockholm, who, on behalf of the King of Spain, was looking for a young naturalist to study the plants of Spain. The young Löfling left for Spain and collected for two years. Shortly thereafter, he sailed to South America, where the climate proved lethal—he succumbed to a fever at only twenty-seven years of age. Similarly, Linnaeus's old companion Christopher Tärnström, a married clergyman with a family, had ambitions to collect in China. He secured free passage on a Swedish East Indian Company ship but got only as far as Indochina, where he caught a tropical fever and died, leaving his widow and children destitute.

The dangers of expeditions, however, did not deter young enthusiasts. Numerous opportunities existed for them because European powers encouraged natural history exploration on account of the potential commercial value of foreign species. European imperialism sought political control to further economic advantages, and the search for natural resources played an important role in European expansion. In naming and arranging products from around the globe, naturalists aided imperial expansion and also implicitly expressed a cultural imperialism. Native peoples might live among a profusion of birds and plants—indeed, the tropics contained a greater diversity than any European country—but from Linnaeus's perspective the local inhabitants were lacking the most basic knowledge. They did not know who created the plants and animals surrounding them, what these objects should properly be called, and how they fit into the established order. According to Linnaeus, the local names possessed no scientific value, nor did they reflect a deeper religious recognition of God's Creation, His Design, or His Will. Just as missionaries attempted to save the souls of indigenous peoples, Linnaeus's apostles sought to save the species of the world for a second naming.

Linnaeus expressed little modesty about his place in this great enterprise.

Adam may have been the first to name God's creatures, but Linnaeus claimed an equally important place. "God has suffered him to peep into his secret cabinet," he wrote, referring to himself in the third person. "God has suffered him to see more of his created work than any mortal before him. God has endowed him with the greatest insight into natural knowledge, greater than any has ever gained. The Lord has been with him, whithersoever he has gone, and has exterminated all his enemies for him, and has made of him a great name, as one of the great ones of the earth."*

Buffon

Linnaeus's main competitor for international preeminence, also born in 1707, outlived him by ten years and was equally significant in the establishment of the modern tradition of natural history. Although they shared a love of nature and a passion for natural history, the two had little else in common. Linnaeus lived the bulk of his career in a small university town, while his rival situated himself in the grandest city of that century, Paris, and served in a highly visible post in the French scientific establishment.

On July 26, 1739, Louis XV of France appointed Georges Louis Leclerc de Buffon, the oldest son of a socially mobile Burgundian family, director of the Royal Garden. The position carried a modest salary and living quarters in the Jardin du roi, as the garden was called. Most important was the prestige and patronage associated with being the head of a royal institution. Buffon, later comte de Buffon, soon became a force to be reckoned with in Paris.

Although politically astute and scientifically informed, Buffon gained his reputation more for his work in the physical than biological sciences. He contributed to introducing Newtonian science in France by translating Isaac Newton's work on the calculus into French, and he entered the Academy of Sciences in 1733 as a member of the mechanics section. The Royal Garden concerned itself with a different set of issues. Louis XIII had established the Jardin du roi in 1635 as a botanical garden for the study of medicinal plants. By Buffon's day, successive directors had expanded its activities. A professional staff gave public lectures on botany, chemistry, and anatomy; gardeners cultivated a wide range of plants; and one of its buildings housed the king's natural history cabinet, or Cabinet du roi.

Although his background did not suggest much expertise in managing such

*Quoted and translated in Knut Hagberg, *Carl Linnaeus* (New York: E. P. Dutton & Co., 1953), 208. The original appears in Elis Maleström and Arvid Uggla, eds., *Vita Caroli Linaei: Carl von Linnés Självbiografier* (Stockholm: Almqvist & Wiksell, 1957), 146.

activities, Buffon harbored some interests in natural history, and fortunately for him they carried considerable political cachet. His work on the strength of wood and the cultivation of forests, for example, proved to be especially relevant. Louis XV's naval minister, the comte de Maurepas, requested that Buffon collaborate with a well-known scientist to investigate problems in reforestation and improving lumber for ships. The research resulted in several publications as well as a successful commercial venture for Buffon. Later, Maurepas was crucial in supporting Buffon for the directorship of the Jardin du roi.

Buffon's career at the Royal Garden turned out brilliantly. He doubled the size of the garden and vastly increased the natural history collection. With Buffon at the helm, the Jardin du roi developed into the foremost institution in its day for the study of the living world.

Buffon's administrative prowess, however, was not the source of his lasting reputation. Instead, his fame rests on an enterprise that he conceived soon after becoming director of the Jardin du roi. Great collections typically had catalogs (which reflected glory on the collection owner), and one of Buffon's first tasks at the royal garden was to produce a catalog of the king's natural history cabinet. Rather than prepare an annotated list of the curiosities and rarities contained in the royal collection, Buffon envisioned a monumental work: a complete natural history of all living beings and minerals. He estimated that the project would take about ten years—a serious underestimation. Buffon would find it necessary to repeatedly revise his timetable. Over a period of almost fifty years, for the remainder of his life, he published thirty-six volumes in which he outlined a theory of the earth and compiled a natural history of humans, minerals, quadrupeds, and birds. (A team of specialists completed the remaining untreated topics during the two decades after his death.)

Buffon's project to write a comprehensive natural history surpassed any earlier attempts. What would be Buffon's resources for such a monumental effort? His education in Dijon, first at a Jesuit college and then at the law faculty, had not included natural history. So, in preparation for his task, Buffon systematically compiled all previous work related to his concerns. He found the ancients—especially Aristotle and Pliny—to be of greater value than more recent authors.

Aristotle, in his *History of Animals,* stressed the value of detailed, firsthand observation, and he collected an impressive amount of information with the goal of uncovering general principles. He assumed that the living world had a general order to it, and, although he did not construct a system to classify that order, he provided many possible starting points for creating one. For Buffon,

The Twinflower

Linnaeus, who named so many plants and animals, has only one plant named in his honor, the *Linnaea borealis,* commonly called the twinflower. It is a surprisingly modest plant to carry such a weighty honor. Linnaeus with a bit of false modesty described it in one of his writings as a lowly insignificant plant, generally disregarded—like himself.

■ *Linnaea borealis,* 1797; author's collection.

Aristotle's writings reinforced his conviction that natural history should be founded on extensive observational knowledge and that it should aim to go beyond particulars to construct an overall picture of the order in nature.

Aristotle supplied an important inspiration, but it was in the writings of his successor, the Roman author Pliny, that Buffon found his model. Since late antiquity readers had respected Pliny as the greatest authority in natural history. In his thirty-seven-book encyclopedia of the natural world, Pliny claimed to have consulted all of the earlier work of Greek and Roman authors. He effectively combined the information to create a comprehensive survey of the natural world: the heavens, the earth, and the animals, plants, and minerals. Individual articles in his encyclopedia that were especially engaging were read by generations of those curious about nature. Later writers added new information and occasionally challenged specific points, but Pliny's status remained high from antiquity through the eighteenth century. Buffon praised Pliny and frequently quoted from his natural history. Like other authors of his time, Buffon was tolerant of Pliny's fabulous tales and seemingly gullible reports—such as that those who gather honey from hives will avoid bee stings if they carry a woodpecker's beak. Such flaws, Buffon reasoned, could easily be corrected. Buffon saved his contempt for authors of the previous two centuries, whom he castigated for gross inaccuracy and mindless compilation.

From the fourteenth through sixteenth centuries, Renaissance humanists sought to supplement ancient Greek and Roman botanical texts with information about plants unknown to Mediterranean authors. Initially, their interest focused on plants of medicinal value, but it soon expanded to include all plants and animals. Their writings were enhanced with realistic woodcuts by Renaissance artists, creating a golden age of nature books in the early sixteenth century. Otto Brunfels's splendid *Living Images of Plants* (1530) is an especially fine example.

Buffon, however, did not appreciate these works. In his opinion Renaissance humanists uncritically gathered all writings about nature without distinguishing reliable observations from fictional or symbolic stories. Ulisse Aldrovandi, for example, published well-known books on natural history. He reproduced fabulous tales and moral lessons as well as reports of investigations he had conducted in his museum. He sought to delight as well as to instruct, and thus also included popular "emblems," a literary and artistic genre of the Renaissance. An emblem generally consisted of a motto, an illustration, and a short poem that was witty or delivered a particular message (such as the value of patience). Since many of these emblems made use of animals, they offered a rich literary tradition from which authors such as Aldrovandi could draw.

Like other Renaissance naturalists, Aldrovandi worked in a profoundly Christian framework; the hand of God, the Creator, could readily be found in all of history, nature, and art. The study of nature led to a natural theology that complemented the revealed theology of Scripture. Buffon's more secular perspective led him to dismiss much of what Renaissance authors wrote as worthless. In his famous discourse on method, placed at the beginning of his *Histoire naturelle,* he stated that if one were to delete all that was irrelevant to the study of nature in Aldrovandi's writings, only one-tenth would remain.

Naturalists in the seventeenth century expanded the observational base of natural history and were more selective in what they included. For Buffon, however, if the study of the living world aspired to be a science—not merely a literary endeavor—an even more rigorous method would be necessary. To set an example, he included in his first fifteen volumes, on the quadrupeds, anatomical descriptions of internal and external characteristics of animals based on specimens in the royal collection. He summarized the most recent knowledge on distribution, breeding habits, life stages, varieties, behavior, and environmental setting, as well as listed the different names given to the animal through the ages by other naturalists.

Buffon patterned the overall structure of the *Histoire naturelle* after Pliny's work, but he dramatically improved its scientific value. Each volume contained engravings to accompany the written descriptions, and interspersed among the detailed articles were general essays that synthesized Buffon's investigations on animal generation, distribution, and classification. Buffon had the advantage over Pliny of having at his disposal a significant natural history collection. He worked assiduously to expand the holdings of the Cabinet du roi, and he succeeded in building it into the greatest collection in Europe at the time. Like Linnaeus, he established a worldwide network of correspondents who sent specimens to the Paris museum, and like his great rival to the north, he also possessed an almost complete library of European literature in natural history.

Although Buffon lacked the knowledge that Chinese and Indian scholars had accumulated, and he dismissed information from the many indigenous peoples the French encountered throughout the globe in the eighteenth century, he nonetheless had resources that dwarfed anything Pliny could have imagined. Buffon's encyclopedia of nature, therefore, reflected a qualitatively different subject matter than found in earlier literature in natural history. Buffon's *Histoire naturelle* helped create a new tradition by presenting detailed studies on a comprehensive scale and by using these studies to attempt to uncover the order in nature.

Buffon's natural history also supplied one of the central documents of the Enlightenment, a new worldview that came into prominence after 1750, first in France and then throughout Europe. The philosophes, the major writers associated with the Enlightenment in France, sought to replace a traditional Christian worldview with a naturalistic one based on human reason. In their attempt to break with the past, the philosophes employed diverse intellectual tools. They used the writings of secular, classical Greek and Roman authors, seventeenth-century skeptics who questioned Christian dogma, and foreign philosophical traditions, particularly English writers who stressed the value of observation. Enlightenment thinkers envisioned new intellectual foundations for government, morality, politics, and art. In their naturalistic worldview, science held a privileged position. They regarded Newton's physical science as the epitome of objective investigation, and writers such as Voltaire popularized the "new English science."

In an attempt to free their contemporaries from Christianity, the philosophes constructed an alternative theological view that depicted God as an abstract geometrician who established matter and the laws of motion but left the system to work out the details on its own. The earth and life sciences could not easily develop out of this deist position. Natural history focused on the particular and stressed diversity. To complicate the issue, theologians who were inclined to tolerate an alternative theological view of the heavens were more

Nature's Caprices

Many naturalists of the eighteenth century described nature as perfect and argued that such perfection reflected God's wisdom and, indeed, proved the existence of a divine Creator. Linnaeus held that God's plan encompassed the appearance of plants and animals, as well as their distribution and relationships.

Buffon, in contrast, argued from a secular perspective and acknowledged the existence of monsters and "less happy" creations in nature. Their existence, he contended, contradicted the religious arguments based on a simplistic notion of nature's perfection. One of his favorite examples of nature's "mistakes" was the toucan. Its beak, Buffon held, was excessively large and impractical.

■   Georges Louis Leclerc, comte de Buffon, *Oeuvres complètes de Buffon* (Brussels: Lejeune, 1828), vol. 13, pl. 127.

conservative when it came to natural history. The theologians of the University of Paris, for example, insisted upon a literal reading of Genesis, and an extensive literature of natural theology had developed which argued that the living world stood as a proof of God's existence and a reflection of his moral laws.

Buffon's encyclopedia supplied a new, secular conception of natural history. Buffon crafted his interpretation in the philosophe style: a clear, popular presentation based on accurate information and understandable to the average educated reader. His articles described nature's wonders, and his essays uncovered its order. As important, his work broke with the Christian tradition that had informed European natural history for two centuries; more accurately, perhaps, he transformed that tradition. For Buffon, like other philosophes, believed in an all-pervasive design in nature. He did not regard that design as the handiwork of a personal Christian God whose truths were to be found in the book of nature as well as in Scripture. Instead, Buffon reified nature into a generative power responsible for the harmony, balance, and fullness of creation. His reinterpretation did not simply stand natural theology on its head by providing a nonreligious interpretation of accepted opinions; rather, it provided a new vision of the living world. Buffon contended that the living world, like the physical world, followed natural laws that investigation could discover. He considered nature an end in itself, however, not a reflection of a higher reality. No suggestion of a Christian Creator or the Christian story of Creation constituted part of his vision.

Buffon even challenged one of the basic premises of natural theology: the perfection of Creation. Although he often described the harmony and beauty in nature, he wrote that "in the middle of the magnificent spectacle" there were "some unheeded productions and some less happy." In his article on the toucan, for example, Buffon explained that nature produces not only monsters like two-headed calves but also monstrous kinds like the toucan, whose beak is "unnatural":

> The true characteristics of nature's errors are disproportion joined to uselessness. All animal parts which are excessive, superabundant, or placed absurdly, and which are at the same time more detrimental than useful, should not be placed in the grand scheme of nature's immediate designs but in the small scheme of its caprices, or if one likes, its mistakes . . . and that whatever proportions, regularity, and symmetry reign ordinarily in all nature's works, the disproportions, the excesses, and the defects demonstrate to us that the extent of its power is not at all limited to those ideas of proportion and regularity to which we would like to fit everything.*

The perfection of nature, then, if one could legitimately speak of it at all, did not consist in the perfection of design or perfection of adaptation. It was not the product of an all-wise and consummate craftsman, who inspired a sense of awe in those who gazed upon his Creation. Rather, the perfection of nature was reflected in the completeness of nature—all that can exist, does.

Buffon's secular vision of nature provided an attractive alternative to Genesis because, in his natural history, Buffon stressed the historical development of Earth and its products. In Buffon's writings, contemporaries found a description of how and when Earth came into being, as well as of the formation of animals, plants, and minerals. Buffon's readers could follow Earth's history from its early molten state to its present stage and could learn the reasons for the current distribution of living forms on the surface of the globe. Buffon explained what animals then existed, how and why they had changed over time, and how fossils had formed.

All of this was described in the *Histoire naturelle* without reference to Scripture or to the direct action of a supernatural power. Instead, Buffon claimed that a basic set of forces, analogous to Newton's concept of gravity, existed and

*Georges Louis Leclerc, comte de Buffon, *Histoire naturelle des oiseaux* (Paris: Imprimerie Royale, 1781), 7:108–9. My translation.

gave rise to animal form and function. These "internal molding forces," as Buffon called them, worked on organic molecules, themselves the result of a chemical evolution on Earth, and thereby led to the diversity of life on the planet. The internal molding forces arose during the early development of the planet. The surrounding environment influenced their expression, and therefore the appearance of the resulting creatures altered over time as animals migrated or as climate and habitat changed. Species that belong to the same "families" would all share the same internal molding force and would be related through descent from an early primitive stock, which arose spontaneously. Dead-end variations left their traces as fossils. Geographical variation resulted from the differing expressions of the internal molding force in different environments.

Like Pliny, Buffon sought to supply his generation with a total picture of nature. He did so in a new fashion: historically. To understand the present, according to Buffon, one had to know the past. If a set of internal molding forces interacted with the environment over time, the key to explaining present-day living forms lay in uncovering the history of life on Earth. This historical dimension of Buffon's science opened a new perspective on life that future generations would develop extensively. It also fit well with a general tendency among the philosophes to explain the present by linking it to the past.

For his contemporaries, Buffon was significant mainly for having written a secular Genesis, which gave the appearance of being grounded on an extensive scientific foundation and on a broad observational base. His scientific peers criticized the underlying speculative elements of his writings, but they also appreciated its boldness. Buffon began to publish his secular creation story in 1749, one year after Montesquieu published his discourse on government, *The Spirit of the Laws.* Diderot and D'Alembert published their monumental project, the *Encyclopédie,* between 1751 and 1772. (It consisted of seventeen volumes of text plus eleven volumes of plates.) The *Encyclopédie* attempted to survey human knowledge from a secular perspective. More than any other document of the French Enlightenment, it served as a manifesto that argued for a rational approach to knowledge and a humane program to change people's thinking and encourage social, intellectual, economic, and political reform. Buffon's *Histoire naturelle, générale et particulière* appeared at a critical period in European thought. His contemporaries regarded it as *the* encyclopedia of the natural world, one that complemented the more general *Encyclopédie.*

The Legacy

Buffon's encyclopedia, combined with Linnaeus's brilliant work in classifying and naming, laid the foundation for the emergence of natural history as a scientific discipline during the second half of the eighteenth century. This is not to say that Buffon and Linnaeus saw themselves as partners. Linnaeus regarded Buffon's flowery prose as a distraction to those who sought knowledge of nature, and Buffon considered Linnaeus's classification systems as little more than boring tables in which to store information. But the combined result of their individual efforts was to set a new level of rigor in investigation, one that gave primary importance to knowledge gained through observation. Nature was seen to operate through natural laws and contained a structure that humans could fathom. The key to understanding nature did not come from Scripture, or contemplation, or mystical insight. It consisted in careful study, comparison, and generalization.

Linnaeus valued naming and classifying. For him, natural history's goal was to construct the catalog of life. The discipline, although based on observation, maintained a deep, religious significance. Many later naturalists who shared his taxonomic bent did so from a wholly secular point of view. In contrast, Buffon placed a secondary value on classification. For him, natural history as a science sought to uncover the broad outlines of the order in nature. That order constituted more than just a list of individual kinds. It portrayed a grand tableau on which natural relationships, driving forces, geographical distribution, and historical change could be recognized. To Buffon, this wondrous picture of nature inspired awe, but he consciously did not conceive of it as connected to the Judeo-Christian story of Creation, or the theological attempts to ground a belief in the existence of God in knowledge of the natural world.

Linnaeus and Buffon thought of themselves as representing different approaches to nature, but they had a lot in common. They each strove for an understanding of the order in nature, and they each chose to conduct their work using a large natural history collection rather than doing their own research out in the field. Museum-based, they valued the arrival of new specimens that would extend the global dimension of natural history. Each formed a network of correspondents to enlarge their collections. Linnaeus and Buffon grasped that much of the globe was still unexplored, leaving them ignorant of much of the planet's richness. They had supplied a foundation, but they knew it would remain for others to complete their project.

2 New Specimens

Transforming Natural History into a Scientific Discipline, 1760–1840

Linnaeus and Buffon established the direction for the development of modern natural history. Together their writings exemplified its goals: to scientifically name, classify, and order plants, animals, and minerals. Their works depended on sizable collections they had labored to build. Buffon had access to the Royal Garden in Paris, which held the largest assortment of quadrupeds, birds, insects, and minerals to be found anywhere in Europe. It also contained a magnificent botanical garden with over 6,000 living plants and a study collection of more than 25,000 pressed plants. Linnaeus's personal collection competed in size and importance with that of the most powerful monarch in the West. When Linnaeus died in 1778 he left 19,000 sheets of pressed plants, 3,200 insects, and 2,500 mineral specimens.

The emphasis given to collecting by Buffon and Linnaeus bore abundant fruit in succeeding generations. Extensive expeditions dramatically expanded the scale of collecting. The resources of the new, vast collections, combined with a set of seemingly unrelated factors (which will be discussed in this chapter), fundamentally altered the study of natural history. The subject expanded, gained rigor, garnered support, and ultimately was transformed into a set of scientific disciplines. The emergence of natural history as a scientific discipline, and its further development into several specialized subdisciplines, can be followed by studying natural history collections—their history, the collectors who made them possible, and the resources that allowed natural history to expand so dramatically.

Collections

Since the Renaissance, when collectors with ties to the court had first assembled natural history museums, natural history helped to define polite culture in Europe. This aristocratic association, combined with the interest of university (mostly medical) scholars in natural specimens, conveyed social status on collection owners. By the eighteenth century, gentlemen considered a modest collection a necessary accouterment, like a carriage or set of silver. The "cabinets" of these amateurs generally contained shells, minerals, ancient coins, and books. Aesthetic considerations guided acquisition as much as anything scientific. As a manual published in 1780 suggested:

> Those who have a considerable number of birds can display them in an enchanting manner by placing them on the branches of an artificial tree, which has been painted green and placed at the back of a grotto-like niche, with a small fountain, in which the water, instead of from a spring comes from a pump or from a small lead cistern placed at roof level to receive rainwater.*

These amateur collections could reach significant size. When Lady Margaret Cavendish Bentinck, the second duchess of Portland, died in 1785, her Portland Museum (which contained one of the most famous examples of Classical art, the Portland vase) took thirty-eight days to sell by auction. Such grand cabinets typically displayed alongside stuffed birds, insects, and minerals an array of other items, including coins, medals, books, and curios. Visitors to a city regarded the cabinets as significant local sights, and contemporary guidebooks noted them proudly. A well-known guide to Paris published in 1787 listed forty-five cabinets worth seeing. Letters of introduction (or personal acquaintance) opened the most exclusive museums to distinguished visitors. A few substantial museums allowed the public in for a fee. Sir Ashton Lever's museum in London, for example, attracted many visitors by emphasizing exotic specimens, such as birds of paradise and flamingos. With his enthusiasm for collecting surpassing his means, Lever charged visitors a fee in an effort to cover his costs.

Exotic specimens captivated the general public's attention, and they provided the most fertile ground for expansion of natural history. This is not to say that naturalists neglected local flora and fauna. The stimulus given by Linnaeus's systematics and Buffon's encyclopedia of natural history inspired numerous local attempts to describe animals, plants, and minerals and to compile these accounts. Some authors followed Linnaeus and provided terse descriptive lists; others, such as Gilbert White, emulated Buffon's engaging prose.

White's *The Natural History and Antiquities of Selborne* (1789) consists of this country clergyman's collection of letters addressed to two naturalists he met in his brother's bookshop. The collection has become a classic of English literature, ranging among the top half dozen most published books in the lan-

---

*A. J. Desallier d'Argenville, *La Conchyliologie ou Histoire naturelle des coquilles . . . Troisième édition par MM. de Favanne de Montcervelle père et fils* (Paris: DeBure, 1780), 193. My translation.

guage. The letters capture in a way that a mere catalog cannot the intimate knowledge of nature that a perceptive rural observer could attain. White's natural history demonstrated the value of carefully studying a small region in detail. As he wrote in one of the letters, "no one man can alone investigate all the works of nature, [and] these partial writers may, each in their department, be more accurate in their discoveries, and freer from errors, than more general writers; and so by degrees may pave the way to [a] universal correct natural history."*

White's work is just one example of the marvelous local floras and faunas constructed at the end of the eighteenth century. By documenting in exquisite detail the natural history of a limited area, they helped raise standards and contributed to a more complete knowledge of the products on Earth.

The general public, although appreciative of the narrower local studies, found exotic specimens more alluring. South American parrots, Australian cockatoos, African orchids, Pacific nautilus shells, and brilliantly colored butterflies from Southeast Asia stood out in collections and fetched handsome prices.

Explorers, colonial officials, traveling naturalists, and employees of commercial houses supplied an eager audience in late-eighteenth- and early-nineteenth-century Europe with specimens of interesting or new species. After the end of the Napoleonic Wars, that steady stream became a torrent as European states began large-scale explorations, marking a period of renewed colonial development. Business, evangelism, and strategic interests motivated this second wave of expansion, whereas settlement had characterized the first wave. Europeans scrutinized every corner of the world in an effort to expand their markets and to save the souls of potential clients.

Natural history occupied an important place in European imperialism. Domination of markets, natives, and nature all went hand in hand. The greater presence of Europeans worldwide and the potential commercial value of many natural products stimulated systematic collecting on a hitherto unimaginable scale, creating opportunities for naturalists to explore exotic regions.

Collectors in the Field

Trained naturalists early in the nineteenth century exploited these new opportunities. They came from a variety of backgrounds. For instance, in 1819 the Paris Muséum d'histoire naturelle established a program to train young travel-

*Gilbert White, *The Natural History and Antiquities of Selborne* (Harmondsworth: Penguin, 1977), 80.

Natural History Collections

Naturalists have been influenced by the environments in which they worked. Natural history collections have been particularly important work sites, serving as centers of research, repositories for reference material, and institutions actively involved in exploration of the globe. These collections have shaped the study of nature in obvious ways, and sometimes more subtly. The emphasis on classification, for example, grows out of the need to order thousands of specimens in a collection; obviously, the study of behavior would be impossible with stuffed birds.

The natural history museum of Paris has been central to the development of natural

history since the eighteenth century. Originally the collection of the king of France housed at his Royal Garden, it became a national collection at the time of the French Revolution and has remained one of the great and defining institutions for the study of natural history. Public lectures based on the collections were given in this amphitheater starting in 1788.

■ Author's photograph.

ing naturalists in collecting, preserving, labeling, and classifying specimens. The program proved successful, and Muséum professors sent trained collectors to areas poorly represented in their collections: West Africa, the Cape of Good Hope, Madagascar, India, Australia, and South America. Other countries soon followed suit, and, before long, representatives from Berlin, Vienna, and Leiden began scouring the remote corners of the colonial world. The situation presented a splendid opportunity for a young man interested in nature and in the exotic to travel and establish himself. It nonetheless carried dangers, and a significant number of these would-be naturalists (like Linnaeus's "apostles") died of disease, accident, or, occasionally, at the hands of local people who suspected darker designs from these intruders claiming merely to be collecting butterflies or seashells for gentlemen in Paris and London.

Even for those who returned in good health, the dream of becoming a self-supporting naturalist often failed to materialize. Young, would-be naturalists faced a daunting struggle to establish themselves. They sought guaranteed employment or funding but were often frustrated. Take the case of Jules Verreaux, whose peers considered him to be one of the century's foremost authorities on birds. He came from a family of taxidermists. His father made a living in Paris stuffing animals and selling natural history specimens; the young man's uncle, Pierre-Antoine Delalande, prepared specimens at the Paris Muséum and had gone on government-sponsored collecting trips to Europe and South America. In 1818 Delalande took the eleven-year-old Jules to the Cape of Good Hope. That three-year expedition resulted in an enormous collection (13,405 specimens; 982 different species of insects alone). Captivated by the beauty of southern Africa and the excitement of observing and collecting, Jules returned to the Cape in 1825 and remained there for thirteen years. He supported himself by sending specimens back to his father's business and by preparing specimens for the museum in Cape Town.

The immense collection that Verreaux assembled during his stay in Africa (along with his scientific notes) would, he hoped, establish a place for him in natural history. But his dreams were dashed when on his return in 1838 the ship, *Lucullus,* crashed on the rocks and sank near La Rochelle on the coast of France. Jules swam ashore, lucky to escape alive. How was he to make a living as a naturalist now? He explored the possibility of returning to the Cape and working in the small museum there but meanwhile sought work at the Muséum, and for a time he helped his father and brothers in the taxidermy business. His luck turned in 1842 when the Muséum appointed him a "traveling-naturalist," to spend five years collecting in Tasmania and Australia. Verreaux succeeded in enriching the museum's holdings while greatly expanding his own knowledge of natural history. Searching for new species and collecting rare specimens, he had the opportunity to observe and record animal behavior in the field.

Upon his return—this time with his collections intact—he supported himself in what became the highly successful Maison Verreaux, which sold specimens to collectors and prepared specimens for display. Eventually he attained a minor position at the Muséum, where he prepared dramatic taxidermic displays (many of them later purchased by the American Museum of Natural History in New York; see Chapter 7). When Verreaux died in 1873 the leading naturalists in Europe mourned the loss of a major authority on birds. Little of Verreaux's knowledge, however, made its way into print because Verreaux had to scramble most of his life to make a living, leaving little time to pursue publication.

The opportunities for a stable career in natural history remained sparse during the early nineteenth century, but the lure of uncovering more of the globe's riches did not diminish. Naval expeditions may have opened up the most exciting avenue to aspiring naturalists. Captain James Cook's three voyages to the Pacific, Antarctica, and Arctic between 1768 and 1778 illustrate the scientific value of major expeditions. Cook's main goals related to astronomy and geography: observing the transit of Venus, searching for the legendary Northwest Passage linking the Atlantic and Pacific Oceans, and establishing the British empire in the southern portions of the globe. His instructions nonetheless included reporting on the natural products he encountered with an eye to any that might be of commercial value. The first two voyages carried experienced naturalists—including Joseph Banks, an enthusiastic botanist and later president of the Royal Society of London, and two of Linnaeus's students—who returned with an enormous number of specimens. Even the tragic third voyage, when Cook lost his life after discovering Hawaii, yielded a substantial number of chests filled with natural history treasures, thanks to one of the ship's doctors (who also did not make it back to England alive).

Even before the end of the Napoleonic Wars the French government mounted several extensive expeditions with specific scientific goals, netting enormous collections. Between 1800 and 1804 Nicolas Baudin led a voyage to Australia on the ships *Géographie* and *Naturaliste,* and he brought back one of the most impressive natural history collections of the early nineteenth century (144 new species of birds alone).

Naturalists or medical officers interested in natural history became a common feature on naval expeditions during the first half of the nineteenth century. Many naturalists who were later to become famous, such as Charles Darwin and Thomas Henry Huxley, gained their first extensive experience on these voyages. Wealthy amateurs also sent out collectors. The German prince Alexander Philip Maximilian of Wied-Neuwied undertook expeditions of his own, visiting Brazil between 1815 and 1817 and North America in the 1830s.

Given the enormous interest in exotic specimens, enterprising naturalists could travel with the hope of selling their collections. William Swainson collected in Brazil at roughly the same time as Prince Maximilian, financing his trip with a modest military pension and some letters of introduction that brought him limited government hospitality. Swainson's life, like that of Jules Verreaux, reflects the new opportunities but also the frustrations that aspiring naturalists faced in the nineteenth century. Swainson's passion for natural history dated from his childhood in Liverpool. His father, a customs officer, collected insects and shells and encouraged his son's interest. The local museum appreciated his talent, for in 1808 they requested that he publish a pamphlet on collecting and preserving specimens. But how could a young man of modest means support himself? Through his father's connections he obtained a military position that allowed him to travel to the Mediterranean. In his spare time there he observed and collected various native flora and fauna. He retired on half-pay with the intention of devoting himself full-time to natural history.

A lucky break in 1816 gave Swainson the chance to travel to Brazil, where he energetically assembled a stunning collection of birds and other specimens. He planned to establish his reputation by publishing an account of the voyage, with descriptions of his specimens. He was sufficiently successful that his publications attracted the attention of the scientific community; the Royal Society of London elected him to membership, and he maintained an active correspondence with many of the leading naturalists of the day.

Reputation, however, did not guarantee a steady salary. To survive financially, Swainson tried publishing illustrated natural history books. Among the first naturalists to experiment with a newly developed printing process, lithography, he produced some strikingly beautiful works. The first was his three-volume *Zoological Illustrations* (1820–23), which artistically depicted shells, insects, and birds. But his income remained irregular and inadequate, especially for a man with a growing family and with some disastrously unfortunate investments. He pursued an opening at the British Museum for which he lobbied using all his scientific contacts, but he did not succeed. The trustees appointed someone who had the proper social contacts but whom Swainson considered unqualified. In exasperation he sold his natural history collections (which included 3,000 birds) and emigrated to New Zealand. Leaving England in 1840, he established a farm in New Zealand and gave up his career as a naturalist.

Swainson's life, like Verreaux's, exemplifies the attractions and frustrations many early naturalists faced. More important, Swainson and Verreaux represent the wave of collectors in the early nineteenth century who aspired to construct natural history on a global scale and who brought new artistic and collecting skills to their work.

The New Collections

European expansion created opportunities for eager naturalists to acquire plants and animals everywhere on Earth. The size, quality, and diversity of the acquisitions made possible a new sort of natural history collection. Until the end of the 1790s most natural history collections belonged to the world of fashion and, with a few notable exceptions, rarely attained appreciable size. Buffon's ambition drove the development of the royal collection in Paris into a major center of research and a repository of materials from French expeditions. The French Revolution threatened the Royal Garden, as it did all the institutions connected with royalty. Owing to the political skill of the garden's supporters, however, in 1793 the revolutionary government accepted a plan that called for the reorganization of the Royal Garden in Paris into the national Muséum d'histoire naturelle. The new public natural history museum housed the old royal collection plus various private collections formerly owned by aristocrats who had fled the revolution. As important, the Muséum offered professorships filled by leading naturalists aided by well-trained staff, who prepared specimens, drew illustrations, and tended the enormous botanical gardens.

The Muséum's success made Paris the center for natural history for several decades. The Muséum served as a model for other new public and semi-public museums in Europe and elsewhere. Only the Dutch came closest to matching the French. Coenraad Jacob Temminck, a highly respected ornithologist, conceived of an imperial Dutch museum that would unite his private collection (one of the largest in the early nineteenth century) with that of the university and with the royal collection. In 1820 the government established the Rijksmuseum van Natuurlijke Historie in Leiden, with Temminck as its first director. Similar to the Paris museum, it served as the repository of the government-sponsored Natuurkundige Commissie, which sent trained collectors to the East Indies.

This merging of collections in Leiden was a trend evident in many other places. As the century progressed, a general coalescence occurred that produced great national natural history collections with trained staff. These were serious working collections, and the museums opened part of their enormous riches to the public. New or revitalized university and private collections supplemented these grand urban collections. By the 1850s even naturalists in what were then relatively remote spots—such as Harvard University in Cambridge, Massachusetts—aspired to build significant collections.

Funding for Naturalists

In the mid eighteenth century the number of individuals studying natural history was small enough that any serious naturalist could correspond with more or less every other like-minded person. By the 1830s, and certainly by the 1850s, the ranks of naturalists had swelled to such an extent that such communication was no longer feasible. The growth of natural history collections and their associated curators were partly responsible for the change. Provincial societies founded numerous museums, many of which were later taken over by municipal governments. These institutions, along with the nascent national collections and university museums, provided employment possibilities for curators. Another source of employment was the private companies that supplied

Popular Natural History

Natural history gained a significant place in the popular literature and mass-market publications of the Victorian age. Authors used nature stories to teach lessons in morality, social behavior, and piety. Publishing houses and philanthropic organizations printed small, inexpensive booklets and pamphlets by the thousands. In addition, popular history became a standard topic for popular magazines. Books and articles on animals and plants served to educate the public, thereby raising the cultural level of the working populace, and had the attraction of encouraging spiritual advancement.

The Penny Magazine of the Society for the Diffusion of Useful Knowledge, which began publishing as a weekly in 1832, had articles on animals in London's Zoological Gardens in its first issue and continued to regularly feature articles on natural history.

This piece on the exotic betel-nut tree typified the popular literature that had such a large audience in the nineteenth-century efforts to broaden knowledge—whether useful or not—and satisfy popular curiosity about the remote and unusual.

natural history specimens, for example, the Maison Verreaux. And, as discussed earlier, there were opportunities for individuals like William Swainson to collect in exotic places with the hope of selling specimens to national, local, or private collections.

Other avenues of funding became significant in many European countries. The Victorian age was one that encouraged "self-improvement," and numerous local societies attempted to facilitate learning among less-educated citizens. Knowledge of natural history was believed to play a significant role in the edification of the "working man," as can be seen by looking at issues of *The Penny Magazine of the Society for the Diffusion of Useful Knowledge* or similar publications that were widely disseminated to the general public. An enterprising person, therefore, increasingly found it possible to secure some financial support by writing popular natural history. Although much of this writing was little more than hackwork, there were authors who took the charge of educating the public seriously and produced thoughtful scientific articles and books for the general reader.

The revolution in the printing industry in the early nineteenth century proved critical for the development of a popular literature in natural history. Manufacturers of printed material took full advantage of the introduction of innovations, such as making paper by means of an endless web on a machine instead of hand-dipping individual pages, applying steam to drive printing presses, and binding books with cloth covers. These changes drastically reduced the cost of publishing and contributed to the enormous expansion of printing at the time.

The invention of lithography was another important advance in the publishing of natural history books, not only because of the process's relatively low cost but also because of the greater accuracy it allowed. In this printing technique, naturalists would draw directly on a flat, specially prepared stone, from which ink impressions were taken to make a printed page. This freed them from their dependence on engravers, who formerly had been employed to copy the naturalists' drawings, onto metal plates, often with a loss of important scientific details. These changes in printing were significant in the development of an extensive popular natural history literature. This literature contributed to extending the public's taste and to opening opportunities for those engaged in producing natural history literature and art.

On a more lavish scale, the "deluxe" natural history book benefited from the public's interest and the new printing technology. The market for these magnificent volumes made possible the creation of such classics as John James Audubon's *The Birds of America* (1827–38) and the forty-five volumes of birds and mammals by John Gould.

Audubon and Gould produced some of the greatest illustrated natural history books of all time. The illegitimate son of a French navy officer, John James Audubon developed as an artist while pursuing a variety of jobs, from dealing in pork and flour on commission in Kentucky to stuffing birds in Cincinnati. Before shooting the birds that would serve as models for his drawings, Audubon made an effort to observe their habitat and behavior. This knowledge, combined with his habit of drawing freshly killed specimens to capture the colors before they faded, gave his paintings added accuracy and scientific value. The desire to observe and draw from fresh specimens led Audubon on extensive wanderings through river valleys, the Great Lakes, and bayous of the South.

Although basically self-taught, Audubon had an extraordinary, romantic style of painting. His striking portraits of birds attracted some of Britain's leading engravers to an ambitious project to publish by subscription a set of large-format illustrations of American birds that could be bound in a series of volumes. Serial publication spread the cost of a work over several years and also allowed a longer time to solicit subscribers. Usually two hundred subscribers would be enough to fund one of these impressive projects.

Robert Havell Jr., the greatest engraver in Britain, produced most of the colored engravings in Audubon's magnificent *Birds of America.* In this "double elephant folio" volume, the large size of the pages allowed the birds to be represented life-sized. When completed in London in 1838, it contained 435 engravings depicting 1,065 birds. The total cost was $1,000. Audubon reissued it in smaller format in the United States (for $100) with text and with less-expensive lithographs replacing the more costly engravings. These smaller volumes quickly became a rich source of hand-colored lithographs (which dealers obtained by tearing apart the books). To this day, used bookstores and print shops sell the drawings as high-status, decorator items.

John Gould, although not an accomplished artist like Audubon, rivaled him in producing illustrated natural history volumes of high quality. Gould made up for a lack of artistic skill with his organizational ability and his good judgment in choosing artists who could transform his rough sketches into watercolors and then into lithographs. Like Verreaux, Gould began his career stuffing animals; unlike his frustrated colleague across the channel, however, he moved from being taxidermist at the Zoological Society of London to become a major figure in the life science community in Great Britain. His success rested on ambitious publishing ventures that joined his business acumen with the skills of various artists and scientists.

Starting with a set of rare birds from India, Gould made sketches from which his wife produced lithographs. He obtained the assistance of the secretary of the Zoological Society in providing accurate scientific descriptions of the birds. Since he could not find any publisher willing to take a risk on an expensive illustrated bird book produced by an unknown author, Gould undertook publication himself, with considerable success. This first work, *A Century of Birds,* published in twenty monthly installments between 1831 and 1832, contained 80 large, hand-colored, lithographic plates depicting 102 birds in life-size. With 335 subscribers, he surpassed the number needed to make a profit and had greater success than several of his better-known competitors.

Gould owed much of his early success to his wife, Elizabeth, who continued to prepare beautiful lithographs for "Gould's bird plates" until her early death in 1841 at the age of thirty-seven. After her death Gould collaborated with other artists on a group of projects that secured his fame as one of the greatest authors of illustrated bird books. He had an eye for what would please the public and a shrewd business sense. In 1849, for example, he began publishing one of his most successful projects, a twenty-five-part series of 360 plates of hummingbirds, the *Monograph of Trochilidae.* The exquisite plates reproduced the iridescent colors of the birds' feathers using gold leaf painted over with transparent oil colors. While the plates were in production, Gould mounted a large display of stuffed hummingbirds and exhibited them for a fee at the Zoological Society. The popularity of the exhibit secured him not only a sizable profit but also additional subscribers. Probably his best-known work, *Birds of Great Britain,* appeared in twenty-five parts between 1862 and 1873. It stands, along with Audubon's *Birds of America,* as one of the high points in illustrated natural histories.

The Specialization of Scientific Disciplines

Although some of the writing in natural history that proliferated in the nineteenth century was aimed at a popular audience, much of it was directed to a narrower audience. The lower costs of publishing allowed naturalists to readily disseminate their new research. With more people engaged in natural history and greatly expanded collections on which to do research, wholly new genres sprang up. The most significant, the monograph—that most standard of modern formats—came into wide use. Monographs offered extensive studies of a single topic and, owing to their inherently limited appeal, could not have become a basic scientific tool without printing being affordable. Similarly, the reduced cost of printing made specialized scientific journals financially feasible. The *Zoological Journal,* one of the first of its kind, devoted itself to animal studies. Journal articles were a means of faster communication in natural history and gave naturalists a forum for discussion of specialized topics.

The rapid growth of research in natural history altered the character of the literature as well as the field itself in other ways. Monographs and articles grew noticeably more specialized, restricting questions of classification to limited subjects, such as a family of birds or genus of plants, and leading to more rigorous standards of analysis. This narrowing of the literature reflected an overall specialization in research. By midcentury natural history had fragmented into separate scientific disciplines and broken into subdisciplines. The first division was into zoology, botany, and geology. These general fields soon gave way to more specific branches. In these new life sciences, the choice of biological group to study was what defined the particular discipline. So, those who studied birds, mosses, or fishes thereby created the disciplines of ornithology, bryology, or ichthyology, respectively. Soon even these categories were too broad to describe the research of naturalists, and more highly specialized subfields arose.

The divisions in natural history gave rise to specialized journals and soci-

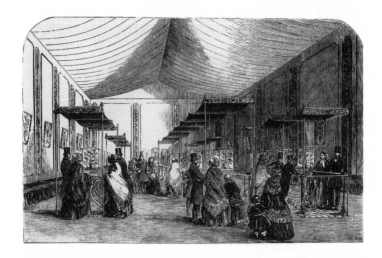

Hummingbird House

Hummingbirds won the rapt attention of the public during the nineteenth century in a way that few other animal groups did. Their small size and insectlike wing movement, combined with iridescent colors in many species, made them a favorite subject for artists and gave taxidermists an opportunity to mount dramatic displays with dozens of specimens. Many a Victorian home had a case of stuffed hummingbirds, and prints of hummingbirds decorated many a parlor. This illustration of the famous hummingbird display constructed by John Gould in the gardens of the Zoological Society, London, reflects the public's fascination with and the commercial potential of these birds.

■ From "Mr. Gould's Collection of Humming-Birds in the Zoological Gardens, Regent's Park," *The Illustrated London News* 20, no. 563 (12 June 1852): 457–58.

eties. Individuals wishing to pursue a career in natural history sought a narrow area in which they could demonstrate their analytical rigor and master the ever-growing empirical information. A leading naturalist of the time, advising those interested in a career in natural history, wrote, "We would recommend to such as really desire to advance [natural history's] progress . . . that they restrict their chief attention to some given department, and when practicable, to those particular groups which have been least studied."* With expeditions to places like Brazil bringing back thousands of specimens—often hundreds of new species of a single group such as ants—the need to specialize made good sense.

*Rev. Leonard Jenyns, "Some Remarks on the Study of Zoology, and on the Present State of the Science," *Magazine of Zoology and Botany* 1(1837):26.

All areas of natural history, however, did not benefit from the push toward ever more specialized research. Some of the best work done in the late eighteenth century and early nineteenth century failed to attract many followers. Perhaps the most striking example can be seen in the legacy of Friedrich Heinrich Alexander von Humboldt. Humboldt had broad interests that led him to travel through the Spanish colonies in the New World between 1799 and 1804, to spend six months on an expedition to Siberia in 1829, and, in the last years of his life, to compose *Cosmos,* an extraordinary five-volume portrait of the physical universe and the history of our knowledge of it.

Although Humboldt's writings influenced a generation of scientists interested in the physical characteristics of our globe, their impact on natural history remained primarily indirect. His engaging writing style inflamed the imaginations of many would-be travelers and inspired Charles Darwin, Alfred Russel Wallace, and Thomas Henry Huxley, to mention only a few of the more famous naturalists who followed in his footsteps.

But of those who followed him, few benefited from the approach to studying nature that he advocated—what historians now term *Humboldtian.* His method emphasized measurement, visual representation, and the search for laws that dealt with complex interrelations. Wherever he traveled he measured and correlated—temperature, pressure, magnetism, moisture, chemical analysis. He recorded the distribution of plants and inquired into their relationships to altitude, temperature, and moisture.

In his *Personal Narrative of Travels to the Equinoctial Regions of the New Continent* (1818–29), Humboldt conceded the importance of taxonomy to natural history, but he argued that science should explore the interactions of organisms as well as their association with the environment. Just naming and classifying would not uncover deeper truths about the living world. Humboldt believed that only by examining what we would today call "ecological" issues (the word did not exist in his day) would we go beyond superficial knowledge of nature. In his sensational and highly publicized ascent of Chimborazo, a mountain peak more than twenty-thousand feet above sea level in the Andes (thought at the time to be the highest mountain in the world), Humboldt noted how the vegetation changed from tropical at its base, to temperate types higher up, and finally to Arctic ones near the peak. He later produced an illustration of the mountain with the different zones of vegetation indicated. Humboldt particularly stressed quantitative measurement and graphic portrayal, which led some field naturalists, astronomers, geographers, and geologists to seek new and improved instruments for the measurement of physical conditions.

For collectors and museum workers, Humboldt's advice had limited practical value. To them, the problems that arose from trying to name and organize the ever-increasing mass of specimens in natural history seemed more pressing. Although they noted the geographic distribution of plants and animals, they had little time to weigh issues they considered to be of only marginal relevance to classification.

Naturalists may have restricted their range of research during the early nineteenth century, but they had not completely lost sight of broader theoretical concerns. The focus on detailed classification raised a number of interesting philosophical and theoretical questions that captured the attention of specialists and the notice of the wider informed public.

3 Comparing Structure

The Key to the Order of Nature, 1789–1848

During the French Revolution of 1830 a friend of the famous German writer Johann Wolfgang von Goethe visited him in Weimar to discuss the violence in Paris. "Now," Goethe exclaimed as the friend entered, "what do you think of this great event? The volcano has come to an eruption; everything is in flames." "A frightful story," the friend replied, "but what else could be expected under such notorious circumstances and with such a ministry, than that matters would end with the expulsion of the royal family?" "We do not appear to understand each other, my good friend," replied Goethe. "I am not speaking of those people at all, but of something entirely different. I am speaking of the contest, of the highest importance for science, between Cuvier and Geoffroy Saint-Hilaire, which has come to an open rupture in the Academy."*

The aged Goethe, like many others interested in natural history, found the heated debates in the weekly meetings of the Paris Academy of Sciences more captivating than the radical political events that had captured most people's attention. Two naturalists, Georges Cuvier and Étienne Geoffroy Saint-Hilaire, championed competing interpretations of the living world. They based their views on material they had studied in the rich Paris museum of natural history. Their disagreement involved two issues. One concerned the basic architecture of the animal body: Did animal form depend upon the functions it performed, or did form reflect a basic "blueprint" that could be understood through comparative anatomy, that is, by comparing the structures in related animals? The other issue pertained to change over time: Have animals over long periods undergone transformations? Various social, political, religious, and intellectual implications gave the debate added resonance.

Comparative Anatomy

The debate in comparative anatomy held significance that extended beyond how the animal body should be understood. Research done in the late eigh-

*Quoted in Toby A. Appel, *The Cuvier-Geoffroy Debate: French Biology in the Decades before Darwin* (Oxford: Oxford University Press, 1987), 1.

teenth and early nineteenth centuries had convinced naturalists that comparative anatomy might hold the key to erecting a natural system of classification
and to the underlying order of nature. The systematic comparison of structure—either structure's basic form or its resulting function—appeared to naturalists as a potentially secure foundation on which to build a scientific discipline to explain the mass of detail known about the living world. Cuvier and
Geoffroy Saint-Hilaire, however, envisioned opposing versions of the new
comparative anatomy.

Étienne Geoffroy Saint-Hilaire had come to Paris as a student just before
the revolution of 1789. He stayed in the capital studying philosophy, law, and
then medicine, but he found his great passion in science, particularly the study
of crystals, an exciting and then-fashionable subject. His career in zoology
came about by chance (as, indeed, did many science careers at this time). In
1792 he had grown concerned about the possible fate of one of his mentors, a
priest whom radical revolutionaries suspected of counterrevolutionary senti-
ments. The priest was imprisoned but later released through Geoffroy Saint-
Hilaire's efforts on his behalf. In gratitude, the priest exerted his influence, and
Geoffroy Saint-Hilaire obtained a post in the Jardin du roi (vacated by a naturalist who left Paris for political reasons). Although only twenty-one years old
and lacking adequate qualifications for the position, Geoffroy Saint-Hilaire
was appointed professor of zoology.

Like other royal institutions, the Royal Garden faced serious threats. The
king had recently been executed, and the ruling powers viewed with considerable hostility everything associated with the old regime. The popular garden,
however, avoided destruction by proposing a plan in 1793 to reorganize itself
into a national museum of natural history (the Muséum d'histoire naturelle).
There, carrying on the Buffon tradition, Geoffroy Saint-Hilaire sought to discover general laws that applied to living beings. He also helped extend the museum collection, accompanying Napoleon on his expedition to Egypt in 1798
and assembling specimens that included two-thousand- to three-thousand-
year-old animal mummies. These materials, he hoped, might shed some light
on the question of whether species had changed over time.

Geoffroy Saint-Hilaire believed that the key to uncovering the order in nature lay in comparing the structures of animals. From his research in comparative anatomy, he argued that he could discern a common plan for all the vertebrates. In his *Philosophie anatomique* (1818), he elaborated on his concept of
"the unity of composition," a basic plan from which one could derive all the
particular vertebrates. According to this perspective, the skeleton and organs
of animals shared a set of structural correspondences independent of the func-

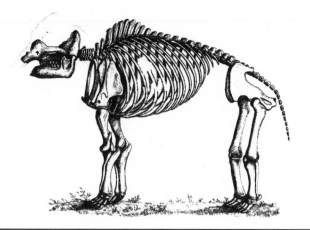

Reconstructions of the Past

The fossil reconstructions done by comparative anatomists like Georges Cuvier elicited excitement and wonder from the nineteenth-century public. Displays of prehistoric animals attracted thousands of visitors to natural history museums, just as today dinosaur exhibits guarantee large admission receipts in modern life science museums. Cuvier pioneered the science of paleontology, and his brilliant reconstructions at the Paris museum of natural history helped popularize the new subject. Large prehistoric animals particularly caught the public's imagination. Cuvier compared the remains of these elephants, hippopotamuses, and other animals to specimens of modern species and demonstrated that they were of different, and extinct, species: an entirely foreign world had existed in the past.

Cuvier used the methods of comparative anatomy for his reconstructions, and his success with fossils gave further credibility to this subject as a key to unlocking the mysteries of nature. One of his most famous reconstructions was of the American fossil *Mastodon*.

■ Georges Cuvier, *Recherches sur les ossemens fossiles de quadrepèdes* (Paris: Deterville, 1812), pl. 5.

---

tions of the structures. His theory received additional support when a disciple, Antoine Étienne Reynaud Augustin Serres, showed that using embryos instead of various adult stages, correspondences in related animals could be established where similarities had previously eluded researchers. Embryological development, in this view, "recapitulated," or repeated, the evolutionary stages of structural change in a series of animals going from simple to complex. "Higher" animals started their development in a simple state and passed through progressively more complex stages as they reached maturity. "Lower" animals did not progress beyond the simple stages, even as adults.

Geoffroy Saint-Hilaire extended his idea of a unified plan to encompass the entire animal kingdom. He concluded that in the geological history of the globe, there had been a set of successive transformations. Jean Baptiste Pierre Antoine de Monet de Lamarck, who also worked on the Muséum collections, agreed. Lamarck studied fossil shells, and his research led him to accept, as did Geoffroy Saint-Hilaire, the idea of the transformation of species. Lamarck believed that two factors drove species change: animals actively adapt to shifts in the environment, and a general progressive force exists in animals that pushes them in successive generations to higher levels of organization. For most of the scientific peers of Geoffroy Saint-Hilaire and Lamarck in Paris, these evolutionary writings seemed too speculative and insufficiently substantiated. The writings consequently attained only limited acceptance in France and elsewhere.

Thus, in the view of Geoffroy Saint-Hilaire, anatomical development stressed form rather than function. Like Buffon, Geoffroy Saint-Hilaire saw animals in a historical framework that included change over time. His colleague at the Muséum, Georges Cuvier, had come to different conclusions. Cuvier grew up on the Swiss border of France and received his education in Stuttgart. Although he had developed a strong interest in natural history (partly from reading Buffon), he hoped that he might secure a position in the administration in Württemburg. When the post did not materialize he supported himself as a tutor in Normandy. The position afforded him some time to study natural history. Soon his goals shifted, and he began to seek a way to pursue a career as a naturalist. He corresponded with various figures of the scientific elite in Paris, and in 1795 he traveled to the capital. There he collaborated with Geoffroy Saint-Hilaire, who aided him in attaining a post at the Paris Muséum. The two naturalists soon parted ways, however, professionally and personally.

Cuvier shared Geoffroy Saint-Hilaire's vision of discovering the laws that regulated the living world, and he believed that comparing the structures of animals would reveal these laws. For Cuvier, however, an understanding of structure had to be grounded in a consideration of the role a particular structure played in the overall animal economy, that is, in its function, instead of focusing on pure form. An organism, he claimed, was a *functional unit* whose structure was determined by its relationship to its overall environment, or its "conditions of existence." By "functional unit" he meant that all organs related functionally to one another and operated together—they could not vary independently of the entire system because a variation in an organ would render it unable to function cooperatively. Cuvier called this broad view of the functional interdependence of organs the "correlation of parts." The animal body, according to Cuvier, displayed intricate integration and exact coordination with its surroundings.

Cuvier believed that through comparative anatomy he could unravel the order in nature. His extensive examination of the anatomy of hundreds of different animals led him to other generalizations. He determined that just as all individuals of a species shared a basic functional plan, so too did the higher-level groups—of genera, of families, and so on. All of these plans, or "types," as he called them, could be rigorously described. Cuvier held that he could organize all the types into a hierarchical system. He presented this system in his comprehensive *Animal Kingdom, divided according to its organization in order to serve as a foundation for the natural history of animals and for an introduction to comparative anatomy* (1817, revised in 1829–30). Unlike most previous classifications, in which animals were arranged in a single, continuous system from simplest to most complex, in Cuvier's system the animal kingdom consisted of four separate divisions without any intermediate links—and without any suggestion of a hierarchy among the four groups. Contemporary naturalists welcomed this classification because of its foundation in comparative anatomy and because of its relative completeness. Cuvier had made extensive use of the collections in Paris to develop his system, and *Animal Kingdom* reflected this.

Cuvier applied his generalizations from comparative anatomy to fossils as well as to living organisms. Indeed, his reconstruction of past life more than anything else caught the public's imagination. In Cuvier's theory, given a single important organ, a researcher could reconstruct an entire animal and describe the environment in which it lived. Cuvier's reconstructions of fossil animals from only partial skeletons demonstrated brilliant paleontological hypotheses, showing the power of comparative anatomy. Many of these reconstructions remain on exhibit in the Paris Muséum.

The examination of the paleontological and geological record did not suggest to Cuvier what it had to Buffon or Geoffroy Saint-Hilaire. How could it? Cuvier's comparative anatomy revealed complex, integrated wholes that seemingly could not be modified without disrupting their functioning. His classification reflected a static picture of nature. To Cuvier the fossil record did not chronicle the ancestral history of contemporary forms but rather presented a history of successive extinctions (except for the remains of the most recent animals and plants). In accounting for the appearance of the fossil record, Cuvier described catastrophes in the geological past which had altered vast regions of the Earth and resulted in mass migrations and mass extinctions. Widely read, his views influenced not only the naturalists' perception of fossils but also the international geological community's ideas on the history of the globe.

Cuvier's zoology had an extraordinary impact on natural history. It combined the taxonomic inclusiveness of Linnaeus with Buffon's quest for the underlying laws of the order in nature. Cuvier produced an intellectually satisfying and scientifically rigorous picture of the living world. His institutional connections—he was permanent secretary of the *première classe* of the Institut de France (the new name and organization of the Academy of Sciences after the Revolution), professor at the Collège de France, and so on—gave his opinions a wider audience and greater authority in Paris than those of his less politically astute colleagues. The centralized nature of French science guaranteed that his ideas were widely disseminated throughout the country. But his reputation extended far beyond France, for Paris remained the international hub of natural history in the first half of the nineteenth century and, therefore, Cuvier's ideas circulated throughout the Western scientific community.

The influence of those ideas reinforced the view that comparative anatomy held the key to understanding living beings. Cuvier's static perspective flatly excluded any evolutionary explanations and focused attention on museum collections rather than field research. He and his contemporaries regarded his science as highly compatible with a religious interpretation of the origin of life. The religious overtones, the strength of Cuvier's conservative political ties, and the centralization of the French scientific elite heightened interest in the famous Cuvier–Geoffroy Saint-Hilaire debate. Although Geoffroy Saint-Hilaire held a professorship at the Paris Muséum, he lacked Cuvier's patronage and political ties. Naturalists on the margin of respectability because of their radical political views or unorthodox scientific opinions saw in the debate a contest between conservative, vested interests and more progressive positions. Social and political biases assumed importance in determining opinion because the issue could not be settled by appealing to observations: the conflict centered on the *interpretation* of the facts.

Embryology

Cuvier and Geoffroy Saint-Hilaire were not alone in their pioneering anatomical studies. During the first decades of the nineteenth century, others in Paris, Germany, and Britain greatly expanded the knowledge of animal form. Most of the studies concentrated on the adult animal, but some moved off into a new area: the examination of embryological development, the formation and stages of life before birth. Investigations of embryo anatomy opened up new avenues in comparative anatomy with significant implications.

Karl Ernst von Baer, an ethnic German from Estonia who spent most of

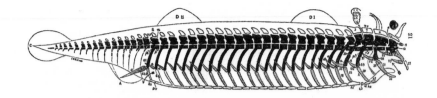

Owen's Archetype

Many comparative anatomists in the nineteenth century believed that the animal body had been constructed according to a basic plan. Georges Cuvier's brilliant anatomical descriptions conveyed the "plan" of hundreds of individual species. Cuvier also argued that more general plans could be sketched for groups of animals, vertebrates for example. British, French, and German life scientists labored to reveal the underlying plans of groups of animals.

Because of its relevance to human anatomy, the vertebrate plan held particular fascination. Richard Owen's *On the*

Archetype and Homologies of the Vertebrate Skeleton (1848) contained a description of the vertebrate plan, the archetype shown here, which impressed many of his contemporaries. Owen assumed that the vertebrate skeleton could be thought of in its simplest terms as a series of ideal units. In the different vertebrates the units have become more specialized. The skull, according to this view, resulted from the fusing of several specialized units, and limbs were merely modified arches of the units.

■   As reproduced in Richard Owen, *On the Anatomy of Vertebrates* (London: Longmans, Green, and Co., 1866).

his scientific career in St. Petersburg, helped to establish the science of embryology. In his classic *History of the Evolution of Life* (1828, the first of three volumes), he detailed the embryological development of the chicken and provided a set of general reflections on the formation of the vertebrate fetus. He believed that the essential nature of the embryo guided its development toward a particular goal or form. Von Baer extended the ideas of earlier writers that development proceeded through differentiation of the three germ layers (i.e., cellular layers—now known as the ectoderm, endoderm, and mesoderm— from which most animal structures originate).

Von Baer explored those relationships among animals which could be gathered from embryology. He criticized the work of Geoffroy Saint-Hilaire's disciple, Serres, and others who accepted the idea that in their development higher animals recapitulated the stages of lower animals. He argued that his more detailed investigations showed an altogether different pattern. Instead of revealing a single continuum from simple to complex (often called the chain of being), von Baer's studies suggested that the animal kingdom displayed four basic types of organization.

Von Baer's four types corresponded roughly to Cuvier's branches of the animal kingdom, and this independent corroboration of a basic fourfold division of animals substantially reinforced confidence in the validity of such a division. Moreover, von Baer claimed that he could detect the type in the early stages of an individual's development. Thus, the earliest stages of all vertebrates shared a common morphology (anatomy), and from there they differentiated to display the characteristics of the more specific group to which they belonged. The traits of the more general taxonomic group always appeared before the more specific group.

Naturalists soon extended von Baer's idea that development proceeded by differentiation, from the more general to the more specific, to claim that embryology could be an aid in solving stubborn problems in classification. A naturalist could determine the taxonomic closeness of two organisms that shared many adult features by searching their embryos for evidence of homologies— that is, for evidence that similar organs developed from the same embryological part. Since homologies could exist only within the same type, according to von Baer, their presence or absence would determine whether similarities in adults resulted from a shared type or were merely superficial resemblances. The classification of animals, therefore, gained a powerful tool for understanding and grouping organisms. Embryology quickly attracted those interested in traditional questions in natural history. For many, embryology represented the forefront of research in uncovering the laws of nature.

The brilliant accomplishments in comparative anatomy by Cuvier, Geoffroy Saint-Hilaire, and von Baer convinced many of the importance of the field. At midcentury, informed observers thought the famous British comparative anatomist Richard Owen the most likely to synthesize knowledge in comparative anatomy into a unified system. Owen had begun his career as curator of the museum of the Royal College of Surgeons, the Hunterian Museum. For comparative anatomy, the museum had few rivals—it housed an enormous collection of skeletons, preserved specimens, and anatomical models of humans and other animals. After twenty-nine years, Owen left the Hunterian to become superintendent of the natural history collection of the British Museum. There he led a successful campaign to construct a separate museum for the rapidly expanding collection (see Chapter 7) and served as the first director of the new British Museum (Natural History). Owen achieved considerable public acclaim for his natural history displays and for naming the group of "terrible lizards," the dinosaurs.

Early in his career Owen adopted a Cuvierian perspective, arguing that the form of an animal had to be understood in terms of the functioning of its organs. But in time he shifted his position and came to argue that comparative anatomy could reveal an understanding of pure form without reference to its function. In his most famous publication, *On the Archetype and Homologies of the Vertebrate Skeleton* (1848), he described the vertebrate archetype. Owen claimed the archetype represented a basic blueprint for all vertebrates. This plan could be discerned by the anatomist in the simplest to the most complex living vertebrates. Similarly, Owen recognized it in the fossil record: the earliest vertebrates in lower strata possessed a primitive version of the plan which differentiated in later strata into higher forms.

Owen's description of the archetype in highly general terms allowed naturalists to interpret the concept in a variety of ways. Some saw it as an ideal plan, a blueprint in a Divine mind that guided the creation of vertebrates. Others considered it the body plan of a primitive ancestral vertebrate from which later vertebrates developed. Owen favored the latter idea, which he believed the fossil record supported, although he could not specify what natural forces operated in the elaboration of the plan.

The broad, popular interest in Owen's archetype reflected comparative anatomy's status in the life sciences in the first half of the nineteenth century. If an understanding of the order of nature could be obtained, Owen seemed poised to provide it through examining vertebrate anatomy.

Early in the 1800s, natural history became more specialized and more rigorous. Museum workers catalogued and named great numbers of plants and animals. Cuvier and an ambitious generation of naturalists aspired to uncover the order of nature through comparative anatomy. But as the Cuvier–Geoffroy Saint-Hilaire debate demonstrated, consensus did not easily emerge on how to characterize that order. Many at midcentury believed that Richard Owen might carry comparative anatomy to new heights in understanding animal form. Natural history had come a long way from the fashionable salon world. Like other scientific disciplines, it possessed a set of defining questions, it had accepted methods of investigation, and it boasted recognized institutions. Well-known naturalists commanded the attention of those far removed from display cases in museums and the primitive camps of collectors in the field. The size, seriousness, and resources of the discipline would soon repay the efforts of so many individuals.

Someone glancing at a portrait of Charles Lucien Bonaparte might easily mistake it for a likeness of his famous uncle, Napoleon Bonaparte. They bore a striking resemblance to each other, and in a curious way the two had much in common. Superficially, of course, they were poles apart: a military genius and famous ruler of France who helped transform nineteenth-century Europe, versus a minor Italian prince who spent most of his life studying birds, primarily in museums. But each Bonaparte ruled as emperor in his own domain, and both men epitomized historical trends.

Not that the two got along particularly well. Charles Lucien Bonaparte had a stormy relationship with many members of his illustrious family. The problems began shortly after his birth, when his father, Lucien, in defiance of his family and his powerful brother's wishes, married Charles Lucien's mother (his mistress at the time) instead of accepting a more expedient political marriage. Lucien thereby legitimized his son's birth but caused a serious breach in the family, and the boy spent his first years in Rome under papal protection. Charles Lucien later repaired some of the family rift by marrying his cousin (a daughter of Napoleon's oldest brother, Joseph). As a consequence Charles Lucien had the opportunity to travel to America to visit his father-in-law (the former king of Spain), who lived during the post-Napoleonic years outside of Philadelphia, in New Jersey.

In the United States, Charles Lucien, who had developed some interest in natural history while in Rome, began a serious study of American birds. He used the Philadelphia Museum as his base. Unlike Audubon, Bonaparte did not explore America's natural history in the field; rather, he spent his time studying collections. Ultimately, he visited every major collection in the world, and by the end of his life he had come close to providing a complete catalog of all Earth's birds. Bonaparte's success represents the work of a set of naturalists who helped bring to maturity projects started by Linnaeus and Buffon. More important, his accomplishments in natural history reflected the resolution of major technical and theoretical issues: how to preserve, display, name, and classify plants and animals; how to standardize those practices; and how to summarize the knowledge so painstakingly achieved by naturalists. These great battles engaged Charles Lucien Bonaparte's generation.

Classification and Nomenclature

The goals of natural history had traditionally been to name, catalog, and order life on Earth and the material products of nature. By the beginning of the nineteenth century the study of minerals had become sufficiently specialized that it branched off as a new discipline, *geology*. Although natural history collections continued to include minerals, and naturalists maintained a lively interest in geology, the distance between studying the material, or physical, world and the living products of Earth steadily grew wider. Increasingly in the popular imagination, natural history focused on "birds and bugs." Ornithology (birds) and entomology (insects) remained important subjects for natural historians, even as they became separate scientific disciplines, as did the study of comparative anatomy. Scientists and the general public continued to regard "natural history" as the collective enterprise of the study of nature's living products (or in the case of fossils, the remains of living products).

But how to settle disputes about the simple naming and classifying of all living things? In 1842 a committee appointed by the British Association for the Advancement of Science to consider nomenclature in zoology observed that naturalists were thrown into confusion over the names of groups of animals even when they agreed on their characteristics. Naturalists in different countries called the same group by different names. According to the committee's report:

> The consequence is, that the so-called commonwealth of science is becoming daily divided into independent states, kept asunder by diversities of language as well as by geographical limits. If an English zoologist, for example, visits the museums and converses with the professors of France, he finds that their *scientific* language is almost as foreign to him as their *vernacular*. Almost every specimen which he examines is labeled by a title which is unknown to him, and he feels that nothing short of a continued residence in that country can make him conversant with her science. If he proceeds thence to Germany or Russia, he is again at a loss: bewildered everywhere amidst the confusion of nomenclature, he returns in despair to his own country and to the museums and books to which he is accustomed.*

---

*"Report of a Committee Appointed 'to Consider of the Rules by Which the Nomenclature of Zoology May be Established on a Uniform and Permanent Basis,'" *Report of the Twelfth Meeting of the British Association for the Advancement of Science* (1842), 106–7.

The rise in the field of natural history of a serious international community that routinely exchanged specimens and traveled to large collections to complete monographic studies encouraged some agreement on the basics. The British committee recommended reforms that would further support such agreement. These reforms included use of the name first given to a group (the law of priority), the use of Linnaeus's binomial nomenclature, and the establishment of the twelfth edition of Linnaeus's *Systema naturae* as the starting point of reference for names. The report also included a set of rules for spelling and for standardizing nomenclature. The clear, well-conceived suggestions were positively received by the zoological community and began a new age of international cooperation.

The naming of plants presented less of a problem than the naming of animals because in 1753 Linnaeus had published a widely accepted list of all known plants and given a binomial name to each in his *Species plantarum*. Later in the nineteenth century the Paris Code of 1867, an international agreement on the nomenclature of plants, formally accepted Linnaeus's *Species plantarum* as the starting point for plant names. International congresses in both zoology and botany continued to examine issues in nomenclature and revise rules on a regular basis.

Technical Innovation

Taxidermy, the display and transport of specimens, and animal and plant illustration underwent their own changes during the first half of the nineteenth century. Although fundamental to natural history, taxidermy—preserving animal specimens—remained a problem throughout the eighteenth century. Without reliable methods of preservation, insects routinely devoured specimens. Dermestes beetles (also known as "museum beetles") could reduce an entire cabinet's worth of exotic birds to mere bones so quickly that the damage was often done without even a hint of the impending destruction. Methods used by many museum curators to try to stem the loss before it was too late unfortunately destroyed the animal skins. Only a few specimens from Buffon's enormous bird collection survive today, most ruined by sulphur fumigations misguidedly used to control insect pests. Finally, in the early nineteenth century, at the Paris Muséum, taxidermist Louis Dufresne discovered that a poisonous arsenical soap developed in the late eighteenth century could keep insects at bay without destroying the animal specimens. His use of the pesticide ensured that his beautiful mounts remained on permanent display, and in his writings Dufresne shared the professional technique with others.

Dufresne's discovery made it possible for taxidermists to focus their atten-

Emperor in a Natural Realm

Napoleon Bonaparte transformed the face of Europe and left a permanent mark on French politics. His influence extended far beyond Europe and had an impact on the United States, the Middle East, and Russia. His nephew, Charles Lucien Bonaparte, shown here (who bore a striking resemblance to the great emperor), sought a career outside politics. During the first half of the nineteenth century, Charles Lucien reigned as the greatest expert in ornithology. His life touched on much more, however. He corresponded with the leading figures in natural history and visited every important collection in the world. His interests extended to the organization of science and its institutions. He played a critical role in establishing the Italian Scientific Congress and worked behind the scenes in promoting and supporting less affluent naturalists. It was Bonaparte who first suggested to Louis Agassiz a trip to America,

and it was Bonaparte who helped Audubon in his quest to publish a work on American birds. His manuscripts in the natural history museum in Paris—fifty boxes of them—reflect the entire range of activities in natural history.

■ Lithograph by J. H. Maguire (1849); author's collection.

tion on the artistic display of birds and mammals. Some curators stored preserved skins in drawers for scientific examination, but most collections open to the public contained naturalistic displays. In Charles Willson Peale's famous museum in Philadelphia—later known as the Philadelphia Museum and so important to Charles Lucien Bonaparte's early career—the backs of display cases had painted landscapes, creating a pleasing backdrop and giving information about each animal's habitat. At the Crystal Palace Exhibition in London in 1851, dramatic, lifelike taxidermic displays created a sensation. A boar being attacked by hounds, and polecats descending on a nest of horned owls ranked among the most popular. Across the English Channel, in Paris, Jules Verreaux, who aspired to be Charles Lucien Bonaparte's successor in ornithology but spent most of his career as a taxidermist, caught the public's imagination at the 1867 Exposition Universelle with his "Arab Courier Attacked by Lions." The American Museum of Natural History in New York later purchased this exhibit, which ultimately made its way to the Carnegie Museum of Natural History in Pittsburgh. Since the last century the exhibit has inspired thousands of schoolchildren (including—sometime ago—me) with a sense of excitement for the exotic.

Perhaps the height of achievement in these artistic displays came by means of the development in natural history museums of dioramas—mounted zoological specimens arranged with natural surroundings, usually against a painted landscape. By the end of the nineteenth-century, biological museums in Stockholm and Uppsala displayed animals and plants against a realistic backdrop to create the illusion of seeing a natural habitat. In the United States, particularly, dioramas became an important part of natural history museums. The American Museum of Natural History in New York and the California Academy of Sciences in San Francisco prepared the most famous ones. Although such displays did not contribute directly to museum research, these popular public exhibits aided museums' ability to attract critical funding to support the overall functioning of the institutions.

Natural history collections, of course, depended upon more than just the ability to preserve specimens or to display them artistically. Transportation of the specimens has always been critical. In the eighteenth century, naturalists might be out in the field for years at a time, with their collections ultimately destroyed by pests or shipwreck. In contrast, nineteenth-century railroads, steamships, canal systems, road building, and harbor construction all meant that more people traveled, goods were cheaper to ship, and shipments followed dependable timetables. Collections thus substantially benefited from the new reliability and efficiency.

The invention of several essential technical devices further increased the scope of natural history in the nineteenth century. Nathaniel Ward invented the aquarium (i.e., the keeping of aquatic animals alive in a case that contained plants in water) and discovered how to maintain living plants in closed glass cases. The "Wardian" cases allowed plants to thrive without being watered while also excluding noxious fumes that would damage or kill them. The cases proved extremely important in providing a protected environment for shipping plants. Salt spray and arid conditions no longer proved fatal to exotic specimens. Wardian cases and aquariums caught the fancy of the public, and soon no parlor, or sitting room, in a middle- or upper-class home was complete without one or both. Bringing live natural history specimens into everyday life also broadened the audience for an already extensive literature on plants and animals.

In an altogether different manner, natural history benefited from improvement in firearms. In the 1820s, the copper percussion cap, containing gunpowder that exploded when struck, eliminated the flash before the shot that had characterized the standard steel spark-and-flint firearms. The flash startled prey, often to their salvation but with disappointing results for the hunter. In fact, the inventor of the percussion principle, the Reverend Alexander Forsyth, claimed his research (which he patented) grew out of his frustration with ducks that escaped his shots because they were alerted by the flash. The percussion cap worked even in damp conditions, including rain.

After midcentury, breechloading shotguns (i.e., firearms designed to be loaded on the part of the gun behind the barrel) with self-contained cartridges extended firearms' accuracy and their safety for gun owners, whether used for sport or for collecting specimens. This was a mixed blessing, for the increasing collections, combined with ongoing destruction of habitats, due to human population growth and economic expansion took a toll on some species. The passenger pigeon, commonly considered so numerous as to be in inexhaustible supply, rapidly declined by the middle of the nineteenth century. By the end of the century, bird watchers reported no passenger pigeons sighted in the wild. Buffalos hit a population low of approximately one thousand before conservation efforts brought them back from near extinction. Modern environmental sensibilities barely existed in the nineteenth century. Naturalists regularly sought the nests of rare birds with the express intention of taking all of them for collections, that is, with the goal of doing so before the species became extinct(!). Audubon, and other artist-naturalists, thought nothing of shooting one hundred specimens in a day—the rarer the species the better. The growth of collections in the nineteenth century may have increased the *appreciation* of nature, but it did little to protect it.

Catalogs and Systems

The development of major collections and the study of the collections by a cadre of scientists not only drove the natural history literature into increasingly specialized channels but also led to what many regarded as the crowning achievement of natural history in the nineteenth century: a set of catalogs that almost fully documented some specific areas. These catalogs reflected the dream of those naturalists who took inspiration from Linnaeus—documenting the entirety of nature.

Charles Lucien Bonaparte's *Survey of the Genera of Birds,* published at midcentury and incomplete because of the author's death, was an attempt to list all known species of birds, over seven thousand at the time. Bonaparte had devoted his life to the study of natural history, and he had the fortune and connections to pursue his interest fully. To draw up his list of birds he visited all the major collections of Europe and corresponded with collectors, museum curators, and naturalists throughout the world. The fifty surviving boxes of his manuscripts and letters testify to the Herculean efforts he exerted in his quest to cover all birds in one catalog.

Other naturalists produced detailed surveys on mammals, fishes, and many smaller groups such as parrots. Curators added to the literature by publishing catalogs of individual museum collections, volumes that soon became major reference tools. Albert Gunther's *Catalogue of the Fishes in the British Museum* was an influential work for a century after it was published between 1859 and 1870. Such catalogs of the great national museum collections listed the sources of individual specimens and other valuable information. Richard Owen, one of the leading British spokespersons for natural history, eloquently stated the importance of museum catalogs when testifying before a parliamentary committee. He claimed that "such a catalogue constitutes, in fact, the soul of the collection."*

The authors of catalogs intended them primarily for the use of professionals or for others who might need the detailed technical information. Compared to the illustrated works of Audubon or Gould, they appeared "dry as dust," but we should keep in mind they had not been written for the broader public. Just as Linnaeus's systematics held little literary appeal, the great catalogs of the nineteenth century possessed a fascination only for initiates in the field of natural history.

Critical Questions

Natural history achieved status as a science through improved research techniques, the increased size of collections, more opportunities for extensive exploration, and its transformation into a cluster of specific scientific disciplines. These factors also inspired public confidence that natural history's goal to name, classify, and order all plants and animals could be achieved. At the same time, however, the very information that provided such hopes also raised new questions and provoked debate.

In Chapter 3, I discussed one of the continuing controversies in the field, the Cuvier–Geoffroy Saint-Hilaire debate. Cuvier, while studying the many interesting fossils uncovered by excavation of the Paris region and acquired by the Paris Muséum, concluded that the geological record reflected a static set of distinct flora and fauna. Others, like Geoffroy Saint-Hilaire and Lamarck, who

*Great Britain, House of Commons, "Report from the Select Committee on the Condition, Management and Affairs of the British Museum," *Parliamentary Papers, 1836*, vol. 2., p. 45.

Artistic Display of Life

International expositions provided many opportunities to bring natural history to the public. Artistic taxidermic displays, which had been developed by a number of private and public museums of natural history, received wide attention at London's Great Exhibition (1851) and the Exposition Universelle in Paris (1867) and led to their increased use. Later in the century, dioramas depicting groups of animals in their natural habitat became standard items in natural history museums. The California Academy of Sciences displays this famous lion diorama.

■ Courtesy of Special Collections, California Academy of Sciences, San Francisco.

also worked on the Muséum fossil collections, came to different conclusions. They held that contemporary species descended from now-extinct ones. Explorations in South America and Australia brought to light fossils that deepened the disagreements and raised new questions.

Examination in the 1830s of the extinct fauna of Brazil and Australia convinced the British comparative anatomist Richard Owen that living beings found in regions of the globe displayed a "localization of type," that is, certain groups of animals and plants, currently and in the past, inhabited only particular locales. What could account for this geographic distribution? Neither climate, soil type, or other geographical factors provided an adequate explanation. Some naturalists attempted to use the problem as an invitation to reconcile science and religion.

Louis Agassiz, a Swiss-born scientist at Harvard University, contended that localization of type represented one dimension of a divine plan that science sought to uncover. Agassiz was strongly influenced by Cuvier, and he accepted the idea that the geological record revealed a history of distinct periods, with each characterized by its own flora and fauna. Agassiz passionately advanced the argument that God had created each species at a particular time and place, a position known as "special creation." The facts of natural history, Agassiz proclaimed, revealed an overall plan. In his *Essay on Classification* (1857) he wrote that by describing that plan, "the human mind is only translating into human language the Divine thoughts expressed in nature in living realities."* Evidence of localization of type demonstrated that the plan not only contained morphological types, physiological functions, and biological interactions, but also had a geographical dimension. God created "South American types" in South America to populate South America, not Australia or Europe.

The localization of type had other implications to naturalists. Recent observations on the current distribution of plants and animals had called attention to the complexity of biogeographical patterns. As discussed in Chapter 2, Alexander von Humboldt pioneered the study of plant distribution, and he had argued that specific environmental conditions had associated "assemblages" of plants. The factors ultimately responsible for such regularities remained unknown, but nonetheless an ever-increasing literature traced the patterns of distribution among plants and animals. Naturalists agreed that the world seemed to be divided into several large regions (generally six), with characteristic floras and faunas.

Naturalists pondered other patterns, for example, the relationship of island flora and fauna to mainland species. Late-eighteenth-century naturalists noticed that the farther an island was from a continent, the fewer species it had in common with that continent. Another pattern that attracted attention was "representative species," in which closely related and morphologically similar species were found to inhabit different geographical areas. Some of the examples struck naturalists as particularly intriguing. In the 1850s Joseph Dalton Hooker, the English botanist and later director of the Royal Botanic Gardens in London, wrote about representative species of plants on the Galápagos Islands. He found it odd that so many plants inhabited only one island of the archipelago, especially when similar but distinct ones could be found on islands elsewhere.

Perhaps the most vexing issue for naturalists in the nineteenth century concerned variation and the "species problem." Large numbers of specimens in

*Louis Agassiz, *Essay on Classification,* in *Contributions to the Natural History of the United States of America* (Boston: Little, Brown, and Co., 1857), 1:135.

museums revealed to naturalists small differences among individuals of the same species. Naturalists came to realize that much of the reportedly significant variation actually resulted from incomplete knowledge of normal sexual, seasonal, and life-stage differences. Some of the variation appeared to be related to geographical location and, therefore, to be constant, warranting the designation of a "variety." How different did specimens need to be to justify the conclusion that they belonged to different species? Although experienced taxonomists claimed they could distinguish species, disagreement about their conclusions often ensued. Buffon's definition of *species*—organisms that could successfully breed together belonged to the same species—did not help because naturalists generally worked with preserved specimens (or descriptions of them) and thus had no way to determine breeding compatibility. According to Cuvier, each taxonomic unit could be rigorously defined through comparative anatomy. Museum workers faced with thousands of beetles or shells found the situation less clear-cut. And so, the "species problem" remained a serious issue.

By the 1850s, natural history was reaching a point of synthesis. Naturalists tackled problems concerning diversity and regularities of life, catalogs such as Bonaparte's listed ever-larger segments of what existed on Earth, and collections housed vast samples from around the world. The general public took a serious interest, as did governments—natural history held religious significance, economic importance, and aesthetic value. Within the scientific community, expectations were high that the subject stood on the verge of realizing the goals its founders had envisioned.

5 Darwin's Synthesis

The Theory of Evolution, 1830–1882

While traveling through Patagonia (now Argentina) in the 1830s, Charles Darwin enjoyed a meal of an ostrich-like bird the expedition's artist had shot. Only after dinner did he realize that he had eaten the bird for which he had been searching. Darwin had been told by locals that in addition to the common rhea, there existed a rare, smaller type. An illustrious French naturalist had recently searched Patagonia in vain for this ostrich-like bird but failed to locate any specimens, and Darwin's sense of competition sharpened his interest in finding the bird. Fortunately, when the cook cleaned the bird he had discarded the head, neck, legs, wings, and much of the skin and larger feathers before preparing dinner, and Darwin salvaged them. Upon his return to England, he sent the remains to John Gould, who identified the bird as a new species and named it *Rhea darwinii.* It turned out to be a significant discovery and would later influence Darwin's thinking about the nature of species. At this point, however, the young Darwin, a recent Cambridge University graduate, spent his time collecting natural history specimens, making observations of native life, and enjoying the exciting trip circling the globe.

Charles Darwin's early career in many ways typifies the training naturalists received during the first three decades of the nineteenth century. Darwin acquired a serious interest in natural history during his university days, first at Edinburgh while studying medicine, and then later at Cambridge, where he went to prepare for the clergy. Because degrees in natural history did not exist at the time, he did not have the opportunity to "train" in that subject, although he did participate in amateur discussion groups and collecting trips. Like other young Englishmen he learned from his elders the moral aspect of natural history, and he later recalled that he had read natural theologians, like William Paley, with considerable pleasure. Earlier in the century Paley popularized this perspective in his classic *Natural Theology* (1802). Given that design in nature reflected a Designer, he argued, a study of organic structures and their intricacies, functions, and adaptation to the environment would prove the existence of a cosmic Planner. Paley's descriptions of the exquisite adaptations that existed in nature, and his conception of nature as a tightly integrated system, particularly resonated with Darwin.

H.M.S. *Beagle*

At Edinburgh, and later Cambridge, Darwin developed strong interests in geology, botany, and zoology. His mentor at Cambridge, the Reverend John Stevens Henslow, encouraged Darwin's curiosity and helped arrange for him to travel on the H.M.S. *Beagle*'s five-year voyage. The British government sent the *Beagle* out shortly after Christmas in 1831 to survey the South American coast and to make longitude measurements around the world. The circumnavigation, in addition to extensive excursions in South America, also stopped at Atlantic and Pacific islands and in New Zealand, Australia, and Africa.

The twenty-two-year-old Darwin made careful observations and amassed enormous natural history collections during the voyage. On his return he parceled out the material to leading specialists in Britain, and then edited and supervised the printing of the descriptions, published between 1839 and 1842 in five volumes as *The Zoology of the Voyage of the Beagle*. Darwin's narrative account of the voyage, which gives an excellent portrayal of field biology at its best, helped establish his reputation. The *Voyage of the Beagle*, as he later titled it, quickly became a popular travel book.

Darwin did more than just collect interesting specimens; he pondered their meaning in light of the central scientific questions that occupied leading naturalists of the day. Patterns of distribution interested him. In South America he encountered fossil remains of giant sloths, rodents, and armadillos, all resembling living forms found only on the South American continent. Like Owen and Agassiz, who studied fossils and extant species from discrete geographical regions, Darwin wondered what accounted for the long-term similarities in characteristics of the living species in a region. Why should there be such striking correlations? What might be the relationship between fossil and extant forms?

Darwin noted other patterns of distribution of living forms while in South America. His geological observations raised the question of how the natural barriers that separated vast areas and their inhabitants had come into existence. He came to realize that the barriers altered with time, sometimes rapidly. He witnessed an earthquake and a volcano and assessed their tremendous power to change the landscape. Whatever the rate of change, natural barriers certainly divided groups of plants and animals in unexpected ways. On one inland expedition Darwin carefully documented the differences in vegetation and animals between the western and eastern slopes of a range in the Andes, differences that had arisen despite identical climate and soil in the two areas.

Darwin took notice of the distribution of closely allied species. He brought back many examples of representative species—among them the plants from the Galápagos Islands which Joseph Dalton Hooker had found so fascinating. Darwin also examined instances of closely allied species that lived in distinct but adjacent mountain ranges. The remains of the rhea (*Rhea darwinii*) that he and his fellow expedition mates had for dinner in Patagonia turned out to be a significant specimen, for the bird inhabited a range that overlapped one where the closely related common rhea (*Rhea americana*) could be found.

The striking diversity of life on oceanic islands and the relationship of the island flora and fauna to that of the closest mainland fascinated Darwin. On the recently acquired British Falkland Islands he observed a wolflike fox that allegedly differed in the East Island and West Island. Later in his voyage he noticed striking differences among the mockingbirds of the Galápagos Islands. Darwin lacked the experience to determine whether his specimens belonged to different varieties or to different species, but the wide variation brought home to him the degree of diversity on these small, isolated islands.

After Darwin returned to England, specialists studied the data he had collected. John Gould, who classified Darwin's avian collection, determined that the mockingbirds gathered on separate islands in the Galápagos archipelago belonged to different species. Why should such close islands have different species of mockingbirds? Equally surprising, Gould determined that the finches Darwin found on various Galápagos islands—as well as another bird that Darwin had taken to be a wren—belonged to a set of closely related but distinct species. The existence on a remote set of islands of over a dozen species of closely allied finches, all of which bore a resemblance to South American ones, puzzled Darwin. Similarly, Richard Owen, in describing the fossil mammals of the *Beagle* voyage, determined that the remains of a llama, although remarkably like the species that currently inhabited the same geographical location, belonged to an extinct llama species. This extreme example of localization led Darwin to ask, why should a similar species replace another in the same environment? His interest in the "replacement" of species was reinforced by Gould's identification of the half-eaten rhea as a distinct species because the two rheas lived in overlapping and quite comparable territories.

Evolution

Soon after his return, Darwin came to believe that if he assumed species had changed over time he could solve all the questions that concerned him. Such a simple assumption would explain issues such as the origin of oceanic life: contemporary species descended from individuals that had accidently traveled from the nearest mainland to the islands. The assumption made sense of distribution patterns: they were the result of dispersion and isolation. It tied fossils by descent to similar contemporary forms. Likewise, it resolved the theoretical problems associated with distinguishing varieties from species by defining varieties as incipient species.

But if species changed over time, how did they do so? What explained the origin of those exquisite adaptations described by Paley? Earlier believers in transformation did not provide much guidance. Darwin knew of Geoffroy Saint-Hilaire's and Lamarck's speculations, but he rejected them as hopelessly vague.

Darwin had certain preconceptions as to what constituted an acceptable explanation of the origin of species. He found untenable the reference to an ideal plan that unfolded in time or to an internal progressive force that caused primitive forms of life to develop into higher ones. These ideas relied on unknowable "forces" or metaphysical theories, whereas Darwin worked in a tradition that expected scientific explanation of phenomena to be formulated in terms of natural laws that governed material objects.

The respected geologist Charles Lyell served as a model and an important influence on Darwin's thinking. Darwin read Lyell's *Principles of Geology* on his voyage, and it determined the way he interpreted most geological facts. Lyell advocated a new geology, one that explicitly rejected "Bible Geology"—the history of Earth conceived of in a theological context—and replaced it with scientific study of lawlike changes on Earth which relied on extensive observation. He rejected concepts such as global catastrophes to explain previous changes on Earth and insisted that geology be based on forces that we can see currently in operation. Inspired by Lyell, Darwin epitomized the new, secular, and specialized natural history characteristic of the early nineteenth century.

Fortunately, Darwin possessed an independent source of income and could, therefore, devote himself to science. Over the next twenty years he constructed a detailed argument consistent with what he knew of natural history. In the process he surveyed the vast terrain of natural history since the time of Linnaeus and Buffon. His interest in variation led him beyond the botanical and zoological literature to journals and studies in horticulture and animal breeding. Through printed questionnaires and personal interviews, Darwin even quizzed breeders who had accumulated an impressive and valuable stock of data.

Sheer quantity of information was not sufficient to establish a theory; what

Darwin needed was a mechanism to explain the change of species. In the Reverend Thomas Malthus's *An Essay on the Principle of Population* (1798), Darwin encountered an idea that he extended by analogy to animals and plants. Malthus had written about human populations, arguing that the natural increase in a human population unchecked by famine, disease, or war would quickly surpass the population's ability to increase food production. The necessary result would be a dramatic struggle for existence. Malthus wrote in the context of the social issues of his day: overpopulation, urban slums, and calls for the reform of welfare. He belonged to a group in England concerned with formulating a new attitude toward the poor, one that was more pragmatic, in keeping with the new industrial society of Britain.

Darwin applied Malthus's idea to the living world in general. Plants and animals possessed enormous reproductive capability. Oysters produced gametes (reproductive cells) that far exceeded the number of oysters found in oyster beds, and oak trees annually produced a prodigious number of acorns. An intense "population pressure" clearly existed in all species. Why, then, should certain individuals survive and leave offspring while others did not? Common sense suggested that survivors possessed an advantage—size, strength, attractiveness, better ability to exploit a food source, resistance to disease, resistance to predators, and so on—and thus produced offspring in greater numbers. A "natural selection," therefore, operated in nature, one analogous to the domestic selection practiced by horticulturists and pigeon fanciers but with a very significant difference: nature lacked a conscious selecting agent. The process had no direction, but over time, and with continued selection, a population might become so altered in character that when compared to the original parent population it might look like a variety, or even a different species.

Natural selection seemed plausible, but did it actually occur? For such a process to take place, considerable variation in individual traits would have to exist. Darwin had accumulated information from breeding and horticulture that suggested there was substantial variation in domestic settings. By chance, Darwin also was engaged in a systematic survey of the barnacles (Cirripedia) because of having collected a highly unusual one during his voyage on the *Beagle*. This investigation led him to appreciate the extent of naturally occurring variation. The barnacle research combined Darwin's interests in distribution, variation, and fossils with traditional taxonomic research stemming from the Linnaean tradition. The work soon grew to enormous proportions, and what started off as a survey to clarify a point in classifying a single specimen became a major research project in classification. Darwin wrote to collectors and collections around the world to request specimens to examine, and over

The Romance of the Tropics

European naturalists found the vegetation and animals of the tropics fascinating. A series of travel accounts in the early nineteenth century, such as those by Alexander von Humboldt, caught the imagination of many young, would-be naturalists and stimulated them to attempt voyages of their own. The result was a second wave of travel literature written by those who succeeded in making exotic trips (and lived to describe them!). Charles Darwin, Alfred Russel Wallace, and Henry Walter Bates are three well-known naturalists whose careers began with expeditions to the tropics. Awed by the beauty and diversity of the living forms, they wrote appreciatively of what they saw.

This illustration accompanied Bates's *The Naturalist on the River Amazons* (London: Murray, 1863). Bates described eleven years of collecting and observing in Brazil between 1848 and 1859, during which time he collected 14,712 species of animals. Some eight thousand of them were new to science.

eight years he studied ten thousand barnacles. It became such a part of his life that according to an often repeated legend, one of his young sons asked a neighboring child where *his* father "did barnacles." In a sense, the project brought together the Linnaean tradition of naming and classifying with the broader search for an order in nature that characterized Buffon's work. Of more immediate importance, by studying multiple specimens of each species of barnacle, Darwin developed a sense of the extensive variation that was possible in nature.

But was that variation unlimited? Cuvier claimed that since all parts of an organism were integrated (the correlation of parts) the variation of any part had a strict limit beyond which it could not go or the organism would become unviable. Darwin, however, argued that the data from natural history suggested an alternative view, for in many cases specimens, or groups of specimens, could be arranged in a series that reflected a gradual change in a particular trait. Varieties within a species and species within a genus could be similarly arranged. This gradation, Darwin believed, reflected a greater plasticity in species than Cuvier had posited. Darwin went further and turned the principle of the correlation of parts on its head by suggesting that changes in one part of an organism would lead to systemic changes in the whole organism, not to death.

Darwin's theory of evolution was revolutionary, and as in other scientific revolutions, an entirely new conceptualization of many fundamental ideas was required. His reinterpretation of Cuvier's principle of the correlation of parts represents one of many basic shifts in perspective. The long-standing tradition of depicting nature as a harmonious balance was replaced with a view that suggested a violent battleground, recalling Alfred Tennyson's phrase, "Nature red in tooth and claw." The "exquisite adaptations" that had delighted Paley could no longer be considered a reflection of divine wisdom and forethought; rather, they were seen as merely fortuitous results of a blind process. Similarly, the complex patterns of distribution, both in time and space, did not represent, as Agassiz believed, the ideas of the Creator but only the consequence of historical operations. Even the concept of species had to be reconstructed. It no longer made sense to think of species as a plan—divine or ideal—but instead species needed to be seen as a population of individuals. Since populations sometimes overlapped, the lines between species became fuzzier.

But if many notions in the life sciences had to be newly conceptualized, an impressive payoff made it worthwhile, for Darwin's grand unifying theory brought together the disparate studies of classification, embryology, behavior, adaptation, morphology, paleontology, and distribution. The observed patterns and regularities of natural history all followed from his explanation of the evolution of living beings. Moreover, his theory resolved natural history's leading questions. Living plants and animals resembled certain fossils because they descended from them, or from closely related ones. Representative species and closely allied species that lived in adjacent areas came from common ancestral populations. The "affinities" recognized in taxonomy revealed historical relationships, and those adaptations that had so impressed Darwin reflected the cumulative result of years of selection of useful variations. The reason naturalists had found varieties so difficult to distinguish from true species became apparent once they realized that varieties consisted of subpopulations that potentially could become new species. Oceanic flora and fauna descended from individuals on the nearest continent, and their diversity of forms reflected successive isolation and the exploitation of new niches. Larger geographic patterns likewise resulted from years of isolation caused by ancient geographic barriers. As Thomas Henry Huxley later claimed, the theory made sense of the "facts of natural history."

Whether working in museums or in the field, concentrating on systematics or morphology, naturalists for the most part quickly accepted the new theory. This did not mean an immediate and radical change of *practice;* a theory may explain facts, provide interpretations of data, and suggest research questions, but it does not necessarily lead to profound alterations in day-to-day activity. Within a decade of the publication of *Origin of Species* (1859), the scientific community largely embraced Darwin's theory, but not without considerable debate. In the process of being adopted, the theory was modified by scientists in significant ways.

Reaction

Although the theory of evolution ultimately gained acceptance, many individuals in the British scientific community initially reacted negatively to Darwin's theory. The recently embraced professional standards of natural history meant that naturalists were uncomfortable with the "speculative" nature of Darwin's book. Many considered collecting and cataloging straightforward empirical enterprises and heaped scorn on attempts to formulate speculative "explanations." Leading scientists and philosophers wrote that science should be "inductive," by which they meant that through repeated direct observations, one could infer valid generalizations. The standard example of good inductive science went something like this: we have observed thousands of swans; they are *all* white; therefore, we infer that swans are white. Darwin had not arrived at his basic concept, natural selection, in that manner, nor did he present it as a process that could be observed directly. Moreover, the cumulative effect of natural selection—change of species—described by his theory operated on a time scale beyond human verification. *If* one accepted the idea of change, then a lot could be explained. But this sort of reasoning did not conform to what philosophers or scientists conventionally meant by "sound inductive reasoning" or good "scientific method."

A set of scientific problems further eroded confidence in Darwin's theory. William Thomson, later Lord Kelvin, raised the issue of the age of Earth. Darwin assumed that Earth was extremely ancient and that natural selection operated so slowly that in the entire recorded history of man no change in species had actually been documented. But Thomson, using contemporary physics—the most exact science of his day and one of impressive mathematical strength—and using accepted ideas on the physical composition of the sun and Earth, convincingly demonstrated that the planet's age could not be as old as Darwin assumed. Thomson also argued from accepted ideas on the nature of the solar system that Earth formerly had been much hotter—too hot to support life—meaning that the time frame in which evolution had unfolded was even more restricted.

This contradiction between physics and natural history upset Darwin and his supporters and stimulated them to consider how the process could have occurred more rapidly. As the twentieth century dawned, the discovery of radioactive elements transformed the debate—uncovering a previously unknown source of energy in the solar system led to renewed scrutiny of the system's age. For several decades before, the seemingly young age of the solar system had stood as a serious scientific problem for those who favored an evolutionary perspective.

Within the natural history community other problems arose. Darwin's theory depended upon the continued inheritance of small variations and their cumulative effect on adaptation. But accepted notions of inheritance suggested that small variations would tend to be overwhelmed in a large population. Naturalists could not see how small, isolated variations could alter a large population: What effect would a droplet of black paint have in a one hundred gallon drum of white paint?

A related issue concerned the extent of variation in nature. Although museum workers and field naturalists had observed a range of variation within populations, they also noted that variation appeared to radiate around a "type." Some corn plants in a population grew taller, some shorter, but season after season agricultural researchers observed the same range in height, even if the seeds selected for the next generation all came from taller plants. Animal and plant breeders had been able to select in some species for desired traits, and to alter the traits' distribution, but there seemed to always be a limit beyond which they could not go (no two-ton zucchini had ever been recorded). And certainly no evidence existed that any new species had ever been produced in all of recorded history. If breeders could not produce new species, how could nature blindly do better? How could an unconscious process like natural selection, working on heritable variation, produce novel forms? Could natural selection produce radical new organs or give rise to the progressive advancement from simple to complex which the fossil record suggested? How could a seemingly aimless force be creative? scientists asked. Where would the "right" variation come from that would adapt plants or animals to complex environmental conditions that themselves changed with time?

Perhaps the most disturbing problems that gave naturalists pause came from comparative anatomy. That discipline's widely accepted laws, as understood by its principal practitioners, flatly contradicted Darwin's directionless evolution that relied on chance variation and natural selection. In Britain the leading anatomist, Richard Owen, although sympathetic to the general idea of progressive development over time of an overall plan, openly criticized Darwin's theory. Owen noted Karl Ernst von Baer's contention that animals could not be arranged into a single series nor could valid homologies be identified among animals belonging to the four different general types. Owen believed that the fossil record reflected progressions from simple to complex within each of the four types. The earliest representatives of a group had the most general features, according to his view, and had differentiated the least. He discerned "archetypes" common to the ancestors and descendants of diverging lineages. In his widely known *On the Nature of Limbs* (1849) he demonstrated in detail how the different vertebrate limbs all shared the same plan.

From Owen's perspective, any discussion of the appearance of life on our planet predicated on chance changes and ignoring fundamental plans missed the most important lessons paleontology and comparative anatomy had to offer. His status within British science severely damaged Darwin's cause because Owen attempted to use his power and rhetorical skills to mobilize the scientific public against the new interpretation of natural history.

Along with scientific doubts, Darwin's contemporaries raised religious objections. To those such as Darwin's former geology teacher at Cambridge, Adam Sedgwick, the *Origin* presented an unacceptable rejection of the most important assumption of natural history: a divine plan gave meaning and purpose to all our observations and joined natural causes to "final" causes. Agassiz had eloquently summarized the facts and patterns of natural history—the very ones that Darwin found so interesting—from this perspective in his *Essay on Classification*. Should such lofty views be replaced with a "rank and unbending materialism?" writers like Sedgwick asked. The issue excited many, even the bishop of Oxford, Samuel Wilberforce, who "episcopally pounded" the Darwin camp.

In spite of serious opposition, Darwin found strong support within the natural history community. Some of those who had traveled on extensive voyages had been led to ask questions similar to Darwin's. Although they had their own interpretations of evolution, the young and energetic naturalists Thomas Henry Huxley and Joseph Dalton Hooker grasped Darwin's originality and recognized the value of his theory.

Huxley and Hooker were joined by Alfred Russel Wallace. Wallace, Darwin's junior by fourteen years and lacking the financial backing and education Darwin enjoyed, had formulated independently a theory of evolution very much like Darwin's. Wallace had traveled in South America between 1848 and 1852, and then in the Malay archipelago between 1854 and 1862. Like Darwin he had been impressed by biogeographical patterns, by the relationship of fossils to living forms, and by the enormous diversity of life. As a young man he found Lamarck's evolutionary ideas intriguing, and he read Malthus.

While in the Malay archipelago Wallace became interested in the origin of species, and in a striking parallel to Darwin, he remembered Malthus's *Essay*. He outlined his own theory of evolution in 1858 and sent it to Darwin. Not surprisingly, Darwin found Wallace's theory to be of interest—he had been working on such a theory for twenty years! Darwin was in a delicate situation. Clearly, he had thought of the theory before Wallace had, and Wallace had not mentioned anything about publication, but Darwin felt obliged as a gentleman to offer to send it to a journal for consideration. His good friend Charles Lyell suggested a fair resolution. He arranged for a reading at the Linnean Society, and for subsequent publication, of Wallace's essay and two short pieces by Darwin which gave them joint credit for the theory. The papers attracted little public attention. Wallace's essay, however, stimulated Darwin to publish the monumental *Origin of Species* in 1859, establishing the seriousness of the theory. Wallace deferred to Darwin's seniority and to the more developed state of his ideas and joined the ranks of supporting naturalists like Huxley and Hooker.

They offered a spirited and effective defense. That so many came to accept Darwin's general idea of evolution so quickly reflected its power to solve many problems, its perceived value as a key to a natural system of classification, and its potential to synthesize the entire field of natural history. For someone like Huxley, it also fit into a broader worldview, one that reflected the secular, industrial society of the nineteenth century.

Numerous writers in the English-speaking world on both sides of the Atlantic went even further and attempted to extend Darwin's ideas to the social realm. This "social Darwinism" generally stressed an analogy between competition in nature and competition in society. Because competition in nature led to "survival of the fittest," many supporters of the social Darwinism argued that human competition would also lead to positive social benefits. Others stressed the inefficiency of nature (thousands of acorns to produce a few oak trees) and contended that humans needed to go beyond biological models and intervene in the social arena. Although Darwin privately agreed with some versions of social Darwinism, he avoided commenting on applications of his ideas to the social realm. All of the formulations rested on highly speculative historical reconstructions of human history or on weak philosophical arguments. Darwin's silence did not impede others, however, from using his name to legitimate social agendas that they claimed followed from his ideas. The continued status of Darwin's name ensured that for the rest of the century, and for much of the twentieth century, social thinkers continued to formulate versions of social Darwinism.

Interpretations

Although naturalists widely accepted Darwin's ideas within a decade, they significantly modified them, typically by incorporating them into a progressive worldview—ironically, one that undercut the main thrust of his writings. The development of evolutionary ideas in the United States best exemplified the "un-Darwinian" way Darwin could be interpreted.

At Harvard, where Louis Agassiz actively campaigned against Darwin, one of America's leading botanists Asa Gray openly supported Darwin's theory and attempted to counter Agassiz's influence. But Gray had his own interpretation of evolution. He accepted the basic argument in *Origin of Species* but considered Darwin's theory incomplete. Gray favored a more religious perspective, one that had a continuity with Paley's earlier natural theology. As long as scientists did not know the origin of the individual variations necessary for natural selection to operate, Gray wrote, we may conclude that God planned and introduced each small variation at the appropriate time and place.

For Gray, far from undermining the idea of an overall Creator in nature, Darwin had extended it by discovering the temporal dimension of the process. Instead of being filled with remorseless struggle and unforeseen consequences, the natural world revealed an overall guided design that led to the natural "creation" of humans by God. Darwinian evolution, as Gray envisioned it, did not contradict traditional religious sentiment. Because Gray emerged as Darwin's leading advocate in the United States—even though he knew Darwin did not approve of his interpretation—his perspective gained a wide hearing.

Other American evolutionists proposed interpretations equally divergent from Darwin's original intent. The fossil evidence as it came to light in the far western parts of the United States gave naturalists an opportunity to propose new evolutionary ideas. American paleontologists detected "progressive trends" in the fossil record and attempted to correlate the trends with corresponding environmental changes. The general progressive appearance of the record, they believed, resulted from a goal-directed force in nature. These Neo-Lamarckians (as they called themselves, in conscious affiliation with the earlier Lamarck)

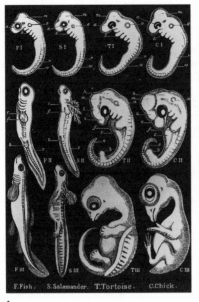

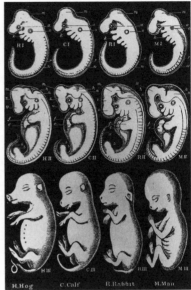

A   B

Ontogeny Recapitulates Phylogeny

Ernst Haeckel drew on the results of rigorous nineteenth-century research in embryology to support Darwin's theory of evolution. One of his most famous discussions centered on the manner in which related organisms followed similar embryological development. Haeckel argued that an organism repeats, or recapitulates, its evolutionary history by passing through developmental stages similar to those of related organisms. His widely read *The Evolution of Man: A Popular Exposition of the Principal Points of Human Ontogeny and Phylogeny* used these plates to demonstrate "the more or less complete agreement, as regards the most important relations of form, between the embryo of Man and that of other Vertebrates in early stages of individual development. This agreement is the more complete, the earlier the period at which the human embryo is compared with those of other Vertebrates. It is retained longer, the more nearly related in descent the respective matured animals are—corresponding to the 'law of the ontogenetic connection of systematically related forms.'"

The plates illustrate the embryos of two lower and two higher vertebrates in three stages of development (*A*) and the embryos of four mammals in the same three stages (*B*). One can see that at the earliest stage all eight are very similar, whereas by the second stage the two lower vertebrates, the two higher vertebrates, and the four mammals have diverged into three groups.

■ Ernst Haeckel, *The Evolution of Man* (New York: Appleton, 1897), vol. 1, pl. VI and VII.

held a wide range of ideas and disagreed considerably among themselves. Some, for example, emphasized the progressive order in nature and the direct action of the environment on development. Others thought trends in the fossil record were nonadaptive, thus indicating developments independent of the environment. University of Pennsylvania paleontologist Edward Drinker Cope, for example, stated that embryological development occasionally accelerated to create new stages of organization. Some of the new stages related directly to environmental pressures: the organism consciously strove to adapt itself by means of its "growth force." Other new stages reflected more formal patterns of development. Cope also integrated his evolutionary views into a broad, religious philosophy. A universal consciousness, he believed, guided evolution and ensured its progress.

In Germany, where Darwin had the greatest contemporary acceptance, naturalists generally incorporated evolutionary ideas into wider synthetic visions. Ernst Haeckel emerged as the major spokesperson for evolution; English translations of his works also attracted a large reading public in England and the United States. Haeckel claimed that Darwin provided the synthesizing vision necessary to construct a complete scientific picture of the living world. By this he meant that the Darwinian theory presented a totally mechanistic vision which did not involve any goal-directed ideas.

Ironically, by pursuing a total system—albeit a physical one—Haeckel often went far beyond "scientific" reasoning. In his writings, such as *General Morphology* (1868) and the more popular works *Natural History of Creation* (1868) and *The Evolution of Man* (1874), Haeckel sketched a highly speculative system that traced evolution from an early, single, spontaneously generated primitive form up through humankind. His use of embryology allowed him to solve one of the major stumbling blocks for Darwin's theory. Since all multicelled animals go through a primitive hollow, two-layered stage (gastrula) in embryonic development, he postulated the existence of an ancestral animal, a "gastraea," which resembled the simple, primitive sponges and from which all higher animals arose. His approach vitiated the objections that comparative anatomists and embryologists had made to a monophyletic (single common stock) picture of evolution. The tradition stemming from both Cuvier and von Baer had stressed the independent nature of four types in the animal kingdom. By positing a common ancestor for the four, Haeckel united the four types. He continued his history of life through higher forms, using comparative anatomy, embryology, paleontology—and his lively imagination.

Haeckel's synthesis partly resolved an old dispute within comparative anatomy between those who stressed the importance of function in understanding form and those who believed that form could be studied on its own terms. From an evolutionary perspective, animal or plant form resulted from many successive adaptations. The general morphological plans that naturalists detected in the plant and animal kingdoms therefore reflected earlier, generalized body plans. Animals and plants, however, also displayed organization that could be directly related to current function. In explaining form, then, the naturalist needed to consider not only the contemporary anatomical and environmental factors but also the legacy of earlier adaptations.

Haeckel argued that it was possible to reconstruct the evolutionary history of life by bringing together information from the different biological sciences, particularly comparative anatomy, systematics, and embryology. In his "biogenetic law" the development of the individual—ontogeny—was seen to recapitulate or repeat, its evolutionary history—its phylogeny. Since evolutionary history followed a series of divergences and branchings, this recapitulation did not resemble a linear movement through a single animal series as envisioned earlier by Serres (see Chapter 3). Instead, Haeckel depicted a phylogenetic tree; that is, a branching rather than a straight ladder.

For Haeckel, embryology held a central place in reconstruction of the past. Researchers could trace and reconstruct ancestral forms by studying the development of living organisms, either to reinforce evidence from the fossil record or, more important, to supply the necessary information where no record existed. His writing on the larval stage of a group of modern marine organisms (ascidians, or sea squirts) provides an example of how he used embryology to fill gaps in the fossil record. These little-studied "worms" have a rudimentary structure resembling one which in vertebrates develops into the vertebrae. Haeckel, following the brilliant work of a Russian embryologist, claimed this shared embryological organ constituted evidence of an ancient link between the vertebrates and invertebrates.

Although Haeckel incorporated some of the best recent research into his synthesis, he mixed in strong social and political opinions. He espoused, for example, a German form of social Darwinism in his widely read book *Riddle of the Universe* (1899). In it he envisioned nothing less than an entire philosophy of life based on evolution. Although Haeckel reached a large popular audience, a number of his scientific colleagues hesitated to accept his overall cosmic picture. Many, nonetheless, agreed with his evolutionary approach, particularly his commitment to evolutionary interpretations of the life sciences.

Whether or not naturalists accepted the sweeping views of Haeckel, by the late nineteenth century they had a common perspective on which to base their ideas about order in nature. Darwin had convinced the scientific world that evolution of living forms had occurred and that change in time explained many of the central issues of natural history. The disagreements over the factors in evolution encouraged research along many different pathways. Comparative anatomy continued to be a major tool of investigation, but new and exciting lines of research soon developed.

6 Studying Function

An Alternative Vision for the Science of Life, 1809–1900

Between 1800 and 1804 Nicolas Baudin led a French naval expedition to Australia and returned with a remarkable collection of natural history specimens. The trip had significance for natural history in other ways as well. Relations between naval officers and civilian scientists on earlier trips had been increasingly acrimonious because scientists often made what captains considered unreasonable requests for space and time, and the strife occasioned by Baudin's expedition served as the final straw. Henceforth, the French government excluded civilian staff and allocated responsibility for scientific observation and collecting to naval officers. In part, attempts to reconcile the conflicting demands of military and scientific interests had exhausted the navy's patience. The decision to suspend the practice of sending scientists on naval expeditions also signaled a shift in priorities. Whereas earlier expeditions had scientific goals high on their agendas, later voyages served more commercial and political ends.

The Baudin expedition represented another significant shift, but one having little to do with naval policy. The expedition brought back botanical substances that were to prove significant for the emerging discipline of physiology, which explores the functions of organisms, including physical and chemical processes. More important, the methods used to study the substances opened up new ways of investigating the phenomena of life, and the success of these methods had profound effects on the practice, status, and future directions of natural history.

Although naturalists emphasized classification and the search for an underlying order in nature, it would be misleading to conclude that they restricted their interest to describing and ordering. Naturalists believed that part of understanding the order in nature was to explore how organisms function. Buffon, for example, constructed a theory of reproduction, or generation, and Linnaeus studied the functioning of the plant and animal body. Other naturalists investigated the processes of generation, nutrition, regeneration, and the excretion of substances from animal and plant surfaces. Understanding the functions of organisms contributed to a deeper appreciation of the laws of nature and a wider comprehension of the order of living beings.

The specialization of natural history which drove research into ever-narrower investigations of taxonomic groups increasingly separated physiological investigation from traditional museum and field studies. In the nineteenth century, studying the functions of organisms became more and more actively centered in the medical community. There it blossomed from a relatively modest enterprise into the exciting and expanding field of physiology. Although the elucidation of function remained of interest to natural history, that field continued to diverge from physiological studies through much of the century. Researchers practiced natural history in museums, government agencies, and some zoology and botany faculties of universities, whereas work in physiology primarily took place in medical settings and in certain zoological or botanical departments and institutes (particularly in Germany). The separation between natural history and physiology had significance for more than just the institutional setting of research; the split led to different approaches to the study of life and to different conceptions of what the unity of the life sciences meant.

Physiology and the Experimental Method

More than anything else, use of the experimental method characterized the difference in the nineteenth century between physiology and other biological disciplines. Since the seventeenth century, investigators in the physical sciences had employed experiments as an important tool, but few in the life sciences had followed their example. The emergence of physiology in France from the 1790s to the 1820s changed that, for the physiologists responsible for building the new scientific discipline shared a commitment to the power of the experimental method.

The development of physiology came about during a general reorganization and revitalization of medicine and surgery in Paris that was part of the institutional reforms initiated by the French Revolution of 1789. Xavier Bichat, one of the principal figures in that reform, held that medical teaching and research needed reorganization and should be based on a new and more scientific medicine. He held further that part of the new medicine should use experimental analysis.

Bichat's experimental surgery to understand the phenomenon of asphyxiation (suffocation) stands as an early example of the power of the experimental method to reveal details of the body's function. Bichat believed that the sudden death of an individual resulted from the initial death of one of three vital organs: the lungs, heart, or brain. Death of one vital organ led to the death of the other two. To demonstrate this, Bichat performed a set of transfusion experiments. By taking the dark blood (deoxygenated) from the lungs of one dog that had experienced lung failure and transfusing it to the heart or brain of another dog, he showed that he could stop the action of each of the other two vital organs in the second dog and thereby bring about that animal's death.

Bichat's early death in 1802, at the age of thirty-one, diminished his direct influence. The task of defining and demonstrating the power of the experimental method fell to his slightly younger contemporary François Magendie. Like Bichat, Magendie brought to physiology elements of his medical background, particularly surgical procedures. To understand the action of different parts of an organism, he operated on living animals, and he successfully established a research tradition in experimental physiology. Starting with a classic paper delivered in 1809, Magendie campaigned to extend the experimental method to the study of life. Magendie's research promoted the experimental study of physiology, as did his editorship starting in 1821 of the first journal dedicated to such study (*Journal of Experimental Physiology and Pathology*).

The story of Magendie's 1809 paper takes us back to the Baudin expedition of a few years earlier. Jean Baptiste Louis Claude Théodore Leschenault de la Tour, the leading botanist on Baudin's staff, fell ill while the ship was in the Indonesian archipelago. He left the ship to convalesce on the island of Java. While there Leschenault learned of two fast-acting poisons of botanical origin which natives used in hunting and warfare. Once he was sufficiently recovered, he brought back to the ship a few samples of the poisons along with some of the plants from which they were made. Most scientific material from Baudin's expedition went to the Muséum d'histoire naturelle in Paris for its naturalists to identify. But Leschenault gave the botanical extracts to two young researchers: François Magendie, who had just obtained his M.D. at the Paris medical faculty, and Magendie's friend and collaborator, Alire Raffeneau-Delile, who along with Geoffroy Saint-Hilaire had participated in Napoleon's expedition to Egypt and was now completing his medical degree.

Magendie and Raffeneau-Delile found puzzling the speed with which one of the poisons worked, and their findings were the subject of Magendie's first publicly delivered scientific papers as well as Raffeneau-Delile's doctoral thesis in medicine (all in 1809). The poison, later identified as strychnine, when placed on a sliver of wood and inserted into the leg of a dog, produced a series of convulsions that ended in the animal's death. The two medical researchers demonstrated that the poison acted on the spinal cord causing violent contractions of the thorax (area between the neck and abdomen) which made respiration impossible and soon resulted in death through asphyxia. Experiments on a horse, six dogs, and three rabbits all showed the same effect. The question was, how did the poison travel to the spinal cord. At that time, physicians were uncertain how the body absorbed foreign substances. Earlier eighteenth-century ideas that the lymphatic system (the system of spaces and vessels in the tissues and organs which circulates a clear, watery liquid, or lymph) served as the major avenue of absorption had been questioned recently by French doctors who argued that the system worked too slowly to absorb the vast amount of material the body must process.

Magendie and Raffeneau-Delile believed that the circulatory system (blood) carried the poison to the spinal cord. The experimental method they used to determine the path of action and to demonstrate it, however, stands out as more historically important than their conclusions. In one dramatic experiment, Magendie anesthetized a dog with opium and then severed one of its thighs from the rest of its body, leaving intact an important artery and vein (the crural artery and vein). He inserted poison into the limb, and the dog quickly showed the expected symptoms and was dead within ten minutes. Because the poison could have reached the animal's torso only through the blood carried by the intact vein, the experiment suggested that absorption of the poison occurred by means of the circulatory system. Some skeptics might have said that very small lymph vessels had remained undetected in the walls of the blood vessel, meaning that the lymphatic system, as argued earlier—not the circulatory system—was the route by which the poison was distributed. So Magendie designed another experiment. He inserted a tube (made from the quill of a bird feather) into the crural artery and vein, tied them off at both ends, and removed all artery and vein in between. The blood then had to pass through the hollow tube to reach the main body, as there were no other avenues for passage left. Magendie obtained results similar to his earlier try.

Ingenious experiments like those of Magendie and Raffeneau-Delile have become a standard part of the life sciences. But in the early nineteenth century they represented pioneering investigations that not only uncovered important information but also demonstrated the power of experimental methods. Because the experiments often involved rather gruesome operations on living animals, we might assume there was a public outcry. Some people did object to the experiments, but public expressions of concern over animal suffering emerged in any significant measure only later in the nineteenth century as humanitarian middle-class values entered the public dialogue. With cockfighting, bullbaiting, and grossly inhumane treatment of animals in markets the norm in the early nineteenth century, animal experimentation did not offend too many people.

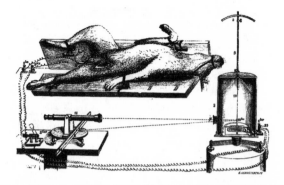

Vivisection and Animal Suffering

The use of experiments in physiology in the early nineteenth century opened up new possibilities in the examination of how the body functions. In France, François Magendie and Claude Bernard pioneered vivisection, operating on living animals to reveal underlying vital processes. This plate illustrates an experiment Bernard performed to measure the temperature of blood in a dog's blood vessels. Although physiologists gained impressive physiological knowledge from vivisections, many members of the public by the 1880s were horrified by the pain and suffering they believed was inflicted on the animals.

The general insensitivity of the biomedical community to animal suffering played into the hands of those who were opposed to vivisection. Frances Power Cobbe in England reproduced illustrations like this one, from a book by Bernard on surgical technique for experimenting, along with quotations from Bernard and other physiologists which suggested scientists needlessly inflicted discomfort on large numbers of animals. The public reacted by demanding government control, which in several countries resulted in laws restricting vivisection and protecting animals from painful experiments.

■ Claude Bernard, *Leçons de physiologie opératoire* (Paris: Baillière, 1879).

Magendie's influence extended beyond the bounds of specific issues in physiology, and his experimental approach influenced early experimental pharmacology and pathology. But with the work of his most famous student, Claude Bernard, experimental physiology fully demonstrated its potential. Bernard's scientific achievements stand among the greatest of the century. He began as an assistant to Magendie at the Collège de France and ultimately followed him, holding the same professorship in that prestigious institution. Bernard's brilliant experimental research earned him the reputation as one of the greatest physiologists of all time. He uncovered the role of the pancreas, clarified the functioning of the liver, and revolutionized the conception of the body. Methodologically, Bernard extended Magendie's use of operating on living animals, or vivisection, by employing a "chemical scalpel," the poison curare (derived from a tropical plant), to block function in an animal at a specific site in the body rather than relying solely on surgical techniques.

In the hands of such skilled scientists the experimental method quickly showed its value in solving difficult problems and encouraged a radically different perspective on the living world. Research in physiology thus strongly influenced the scientific perspective at the time. In his highly popular *Introduction to the Study of Experimental Medicine* (1865), Bernard described all bodily phenomena as resulting from specific causes and labeled this approach "determinism." He argued that determinism should become the guiding principle in physiology, meaning that researchers should regard the body as a mechanism that could be understood using chemical and physical tools. The body, then, resembled the solar system, that is, it was a system whose regularities could be discovered, and physiology potentially was a rigorous science like physics.

Bernard also broke with centuries of medical tradition in which function was understood by reference to specific organs. His "general physiology" instead proposed that functions resulted from an integrated system, not a single organ. Of broader significance, since the study of functions that characterize all living beings belonged to the province of physiology, Bernard believed this study—or "biology," as Auguste Comte, an influential French philosopher of the early nineteenth century, termed it—would uncover the laws of life.

Biology was a term used since the beginning of the 1800s to characterize a unified view of living organisms which stood in contrast to the general picture constructed by natural history. Instead of searching for order in the diversity of animal and plant form, or in the overall classification of living beings, biology sought what it took to be more fundamental laws that characterized the basic functions of all organisms, for example, nutrition. Because of its association with the experimental method, these laws allegedly were more rigorous and more like the physical sciences than natural history. New societies, like the Société de Biologie to which Bernard belonged, had as their goal a new science of life, one that would transcend the applied world of medicine and the old, dusty one of natural history. The stunning success of physiological research raised hopes that the new knowledge would provide a foundation for an effective medicine and a synthesizing theory of life.

The development of experimental physiology soon spread to other countries. The growth of research in Germany, in particular, had lasting and important consequences. Most noteworthy the "1847 Group," a set of young researchers, aimed to understand life in purely mechanical and material terms, that is, they aspired to explain biological function in terms of physics. Their rigorously analytical investigations showed the value of the techniques of the physical sciences and of new technological instruments in understanding the body. Later on, they realized that their youthful enthusiasm to completely reduce life to a branch of physics had been overly ambitious and somewhat unrealistic, but the quantity and quality of their individual research nonetheless stands as a remarkable achievement. By the middle of the century, partly because of the institutionalization of this laboratory science in the new university system of Germany, the volume of physiological research grew strikingly. After midcentury, it continued to expand in Germany and in the established universities of France and Britain.

Physiological research significantly affected the overall conception of the body. Earlier in the century German naturalists and philosophers had speculated about the existence of a basic unit of life. Scientists in Germany who held the mechanical and material approach of medical physiology took that idea and, using technological innovations—such as lenses for the microscope that did not distort the image—elaborated a view of the body based on an elemental unit, the cell. Matthias Jacob Schleiden using plant material, and Theodor Schwann using animal examples, outlined the basic cell theory: the body could be understood as an organized system of cells, and vital functions resulted from chemical reactions occurring in cells. The high point in the theory's development came the year before Darwin's *Origin* appeared, with publication of *Cellular Pathology* (1858) by the German scientist Rudolf Virchow. This collection of lectures (which originally had been delivered in Berlin that same year) set out a complete picture of the body, in both health and disease, in terms of cells and cell activity. Virchow's monumental synthesis helped establish the conviction among many life scientists that the cellular level was the proper level at which to generalize about the living world. For Virchow and others, the aim of life scientists should be to uncover the laws of life, rather than "old-fashioned" classification and speculative searches for an order in nature.

The French and German approaches to physiology differed somewhat in their basic assumptions about life. Many German scientists accepted a materialist philosophy, which viewed all life's functions as resulting from mechanical and material causes, while their French counterparts more typically held doubts about our ability to know the "ultimate" nature of life. But German and French physiologists had more in common than not, and all held a strong commitment to the experimental method. They also shared an ambivalence about Darwin's theory of evolution and the claim that it constituted a synthesis of our knowledge about the living world.

The physiological tradition had a different orientation that made it difficult for those in it to appreciate Darwin's achievement. For French laboratory
scientists such as Claude Bernard, Darwin's theory lacked scientific demon-
stration. Physiologists, who focused on the current activity of vital functions,
had no way to evaluate an argument that dealt with the history of life on Earth.
The continuing influence of Cuvier in natural history, combined with physiologists' skepticism, ensured that Darwin had little impact on French science
in the nineteenth century.

German materialists, who saw in their rigorous, analytical investigations
the path to progress in understanding life, considered Darwin's theory a back-
ward trend toward the speculative notions of earlier times. Even those sympathetic to the idea of the incremental development of life over time, like Virchow, found Darwin's level of synthesis inappropriate; the great synthesis of
the study of life appeared more likely to come at the cellular level than in grand
visions of the history of life, most of which would never be known in any detailed fashion.

Physiology and Natural History

The physiological tradition has significance for natural history in ways more important than just an initial hostility to Darwin's theory. For it represented a competing tradition in the life sciences that not only held its perspective to be the
more valuable one in shedding light on the nature of life but also claimed to
deserve a larger share of the institutional resources because of its superior intellectual claims and its potential practical value for medicine and agriculture.

Consider, for example, the rivalry between physiology and natural history
in Paris in the late nineteenth century. The Muséum d'histoire naturelle was
the leading institution in the biological sciences during the first half of the century. As discussed in Chapter 2, it had been reorganized during the French Revolution from the old Royal Garden into the new national museum of natural
history. In place of an autocratic "Royal Intendent," who had comprehensive
powers in managing the Royal Garden, a less powerful director, elected one
term at a time by the professors, led the Muséum. Of more consequence, in
place of a curator of the royal collection and a set of lecturers in botany, chemistry, and anatomy, the Muséum established twelve "professor-administrators."
The areas of study of these professors reflected the new specialization that
would soon characterize all natural history museums. In time, the museum created separate professorships to deal with ever more specialized branches.

The history of the Muséum positions reflects the battle between physiology
and natural history. Until 1837 new professorships were created as the collections

in natural history expanded. But a shift toward experimental work was reflected in those professorships created in the four decades that followed. During the 1880s and 1890s the museum added seven professors in experimental subjects and ten associated with natural history collections, although even among those ten some researchers conducted experiments. The very center of the naturalists' kingdom appeared to be poised for a possible takeover by "biologists."

The status of natural history relative to experimental biology as shown by interinstitutional rivalries reflects an even more telling set of events. Until the middle of the century, the Muséum reigned as the central institution in Paris devoted to the study of the life sciences. Beginning in the 1850s, at the same time as the challenge to natural history, the Muséum faced increased competition owing to the development of programs in physiology at the Collège de France and at the Faculty of Sciences at the University of Paris. The Faculty of Sciences established the first professorship in physiology in Paris. As the century progressed, the Faculty of Sciences grew in size and importance, ultimately overtaking the Muséum, and contributed to changes in scientific patronage and career patterns.

The Muséum professors mounted a revival of natural history toward the end of the century by trying to play a role in the expansion of French colonies. Professors aided the identification of native flora and fauna and gave advice on introducing new plants and animals for farming. The effort achieved only partial success. It shifted the power within the museum back into the hands of the naturalists, but at a heavy price. Although the Muséum increasingly left physiology to the Collège de France and the Faculty of Sciences, it failed to assert its own importance in the colonial enterprise.

From an intellectual vantage point, physiology and natural history did not have to be competitors. During the eighteenth century naturalists felt no tension between the study of the order of nature and the investigation of the functioning of the body. In the nineteenth century Haeckel's brilliant synthetic vision integrated Darwin's theory with the physiological tradition in Germany (see Chapter 5). Haeckel argued that physiology could explain the functioning of the individual, and comparative anatomy could uncover patterns of form, but that neither provided a complete view because neither could explain the origin of form and function. For Haeckel, Darwin had supplied that piece of the story. The coalition of ideas forged by Haeckel, however, achieved only limited acceptance (mostly in Germany).

Potential Synthesis

Although the traditions of natural history and physiology did not have to be in opposition, they nonetheless often were. Competition for resources accentuated the conflict. Toward the end of the nineteenth century changes in the life sciences raised the possibility of an integration of the separate traditions.

Impressed by the rise of experimental physiology and its spread throughout the scientific world from Baltimore to Kiev, naturalists sought to import the experimental method into areas of natural history where certain questions appeared intractable or where new questions could be profitably asked. Over a period of a few decades a revolution in the biological sciences took place. Institutional changes reinforced intellectual trends, and in spite of the disarray in the study of evolution by the end of the century, there were research results that made the construction of an integrated picture of the living world seem more achievable. In the study of heredity and embryology, particularly, the most dramatic developments occurred.

From the time of von Baer, embryology had been important in natural history and maintained a central place in discussions of classification and comparative anatomy. Cell theory, which sought to explain the phenomena of life in terms of cell activity aided by technical advances in microscopes and microscopic techniques, refined the conceptual and physical tools available for research. The use of experimental methods in embryology vastly extended the discipline's power. Anton Dohrn, a German naturalist, had a significant part in introducing these methods by establishing the Naples Zoological Station in 1872. A former student of Haeckel's, Dohrn set up the Zoological Station to facilitate year-round research on animals from the warm waters of southern Italy.

Starting in the 1880s, investigators at the Zoological Station began experimental studies to clarify certain issues in embryology. Wilhelm Roux and Hans Driesch conducted the most famous of them. The two scientists worked primarily to assess the extent to which the embryo operated as a self-differentiating system, as opposed to a self-regulative one that responded to environmental conditions. Roux argued, on the basis of a set of classic experiments, that the developing embryo's cells differentiated, that is, became specialized, owing to internal factors, not in response to an external stimulus.

In one experiment, Roux rotated eggs so that their heavier ends were up, not down, and showed that they developed normally, thereby, he believed, refuting the idea that the orientation of the frog egg constituted a determining factor in its development. In another experiment, he punctured and destroyed one of the two cells (blastomeres) after the first cleavage in the frog eggs and observed abnormal surviving embryos—they developed into partial versions of the blastula or the next (gastrula) stages. That is, the surviving blastomere developed as it would have had its destroyed partner continued to develop.

Roux used such experiments to argue that development did not consist in an interaction of cells among themselves and their environment, as many believed, but rather that embryological development followed a qualitative differentiation, or specialization, where cells received certain particles from their parent cells which determined their fate. This "mosaic" theory of development suggested that in time researchers would be able to follow the history of each specialized cell from the initial fertilized egg.

Roux helped popularize the use of experimental methods to resolve embryological questions, a field of study he called *developmental mechanics,* the search for the physical and chemical causes of the development of form. The new field stressed asking interesting questions for which tests could be devised to provide answers. In contrast to fruitless speculation, this approach promised to help scientists reach valid conclusions.

Although a powerful tool, the experimental approach of developmental mechanics did not prove a magic elixir. This can be seen by considering the research of Hans Driesch, whose experiments contradicted Roux's conclusions. In one set of experiments parallel to Roux's, Driesch used sea urchin eggs, separating the two blastomeres rather than destroying one. He found that each of the two blastomeres gave rise to a half-sized blastula. Each cell appeared to have the potential to develop further. Driesch concluded that the developing embryo revealed a more plastic nature than Roux believed and that it responded to its environment. This contradicted Roux's mosaic view that cell division resulted in hereditary material splitting unevenly with the result that cells became increasingly specialized because of their differing hereditary material. Driesch did not accept the purely material explanation that Roux stressed, and he later championed a "vitalist" interpretation, one based on an entelechy, a vital agent or force, that guided organic development. A rather heated debate ensued.

Although Roux and Driesch continued to disagree, their research demonstrated to others the experimental method's value in probing nature. Naturalists at the Naples laboratory studied the early cleavage stages in development, and visiting scientists from around the world caught the excitement these studies generated. The Americans E. B. Wilson and Thomas Hunt Morgan, for example, studied at Naples and brought back techniques they learned. The new studies complemented their observational skills. Wilson and his colleagues extended the search for homologies (structures with the same embryological origin), a traditional subject in morphology, all the way to the cell through a pioneering set of studies on cell lineage. These cell lineage studies traced the history of specialized cells back to the earliest divisions of the egg. Using new methods to determine homologies in the very early stages of embryological de-

Biology Lab

The introduction of laboratories as part of instruction
in the life sciences in universities improved teaching
and became standard on
both sides of the Atlantic.
The model spread widely.
This photograph of a teaching lab at Oregon Agricultural College (later Oregon
State University) dates from
the 1890s.

■ Courtesy of the Oregon State University Archives (P25:5).

velopment took the study of the germ layer another level deeper. Other fields
followed a similar pattern as the integration of experimental techniques with
the observational practices of natural history broadened the range of options
open to investigators.

As in embryology the study of heredity, which sought to explain the trans-
mission of characteristics from parents to offspring, underwent major changes
because of the knowledge and ideas generated by experimental methods. In
light of Darwin's theory of evolution, which depended upon there being a
means by which to transmit individual variation from one generation to another, scientists conducted research on heredity. Darwin had devoted a two-
volume work to the subject—the *Variation of Animals and Plants under Domestication* (1868)—where he formulated a theory of generation compatible
with an evolutionary perspective. His "pangenesis" hypothesis explained generation by speculating the body produced *gemmules* throughout the individual's life cycle. These hypothetical particles of heredity ultimately created the
male and female contributions to reproduction. The idea of gemmules allowed
Darwin to explain how individual variation could be transmitted and to incorporate the notion that characteristics acquired during the life of an organism could be inherited, thereby speeding the process of adaptation. Many scientists attempted to extend the hypothesis, but with little success.

Cell theory initially disagreed with Darwin's hypothesis that the smallest
units of the body threw off gemmules that collected in the seminal fluid. It assumed that cells came from preexisting cells, not from particles thrown off
from cells. Roux's embryology, which assumed a qualitative division of mate-

rial constituents of the cell, did not fit with pangenesis and neither did the embryological views that stressed the importance of the environment in which development occurred. Cell theory did, however, turn out to support the idea of pangenesis. August Weismann, one of Darwin's chief supporters in Germany, worked out an important theory that incorporated the knowledge gained in cytology—the study of cells—into an evolutionary perspective. His theory, moreover, helped reorient the study of inheritance.

Weismann built on exciting discoveries and experiments in the study of the cell, and in particular on fertilization. He made a basic distinction between the cells that made up most of the body, the soma cells, and those that contained the hereditary material, the germ plasm. The soma cells did not influence the germ plasm, and therefore any changes in the body acquired during the life of the organism could not be transmitted. The germ plasm, Weismann claimed, contained complex chemicals called *biophors,* which regulated the form and function of each cell. The biophors in turn were organized into units called *determinants,* which controlled cells or groups of cells, and the determinants were joined into *ids,* which could organize an entire body.

In Weismann's theory, a physical continuity in the hereditary material was said to persist from generation to generation and this material consisted of discrete units. Weismann believed his theory supported Darwinian evolution. The mixing of hereditary material from both parents ensured a large pool of variation on which selection could operate. Although other theories ultimately prevailed, Weismann's emphasis on discrete, material, hereditary units remained of fundamental importance in reorienting the study of inheritance.

Thus, by the end of the nineteenth century a curious situation existed in the life sciences. What had been separate traditions, natural history and physiology, moved toward a synthesis. The experimental method that had been the hallmark of physiology no longer uniquely characterized the study of function and had become an important avenue for exploring many traditional topics in natural history. Scientists increasingly discussed the experimental investigation of evolution, morphology, inheritance, and development, but a general synthesis remained elusive.

In large part, the failure to develop a synthetic treatment of the life sciences reflected the problematic state of the theory most likely to provide that synthesis, the theory of evolution. Darwin had won the war but lost his most important battle. Although the biological community widely accepted the idea of evolution, and many in the life sciences saw it as the unifying thread in understanding the living world, there was no agreement on how it operated. Darwin's reliance on natural selection found few supporters. Vernon Kellogg, an

American naturalist sympathetic to the Darwinian approach, described the consternation of those around him at the turn of the century in *Darwinism To-Day: A Discussion of Present-Day Scientific Criticism of the Darwinian Selection Theories, together with a Brief Account of the Principal other Proposed Auxiliary and Alternative Theories of Species-Forming* (1907), which reviewed the widely diverse alternative evolutionary positions. The increase of knowledge in different branches of the biological sciences seemed to make the situation worse rather than better.

The lack of synthesis in the life sciences did not mean the disciplines remained static. Large fields of study realigned themselves, with significant implications for the naturalist tradition. Specialization had led many researchers to define themselves by new disciplinary subjects rather than in traditional terms. As cytology, embryology, and the study of heredity became more institutionalized, the older category of "natural history" and the designation of "naturalist" began to shift in meaning. Those researchers who used the experimental method and, typically, worked in laboratories, institutes, and university departments rejected the "old-fashioned" label *natural history* and used newer terms that indicated their specialized field (e.g., *embryology*) or an alternative general name such as *biology. Natural history* and *naturalist* referred to work and workers in collections or in the field. Natural history came to be associated with systematics, evolutionary morphology (i.e., the construction of phylogenies, evolutionary histories), and the study of distribution.

To make the situation even more confusing, educators no longer used the term *biology* as Bernard and Comte had—to connote a physiological approach to the living world as opposed to a natural history one. Thomas Henry Huxley, for example, had proposed as early as the 1870s that science and medical education encompass the entire study of life. He called this broad subject *biology,* and he believed it should examine everything from the cell to evolution. Huxley labored to create new educational institutions (and reform old ones) which would stress science in their curriculum. His work at the Royal School of Mines, where he led its expansion into a school of science and a training institution for science teachers, significantly affected science education. As dean and professor of biology, Huxley emphasized a hands-on, laboratory-style biology education. His influence on the English curriculum profoundly affected the leading institutions of Britain.

Huxley called for an expansion of the teaching of science and for its reorganization. He particularly insisted that "biology" be taught in schools. *Natural history* struck Huxley as an outmoded term that had been used by so many people in so many different ways it should be replaced. The vast specialization in natural history had yielded too many separate branches: the study of birds, fish, minerals, plants, and so on. Moreover, groups of scientists used different approaches in their study of those living beings: for example, morphology, physiology, and embryology. Huxley wanted a new term to characterize the study of the totality of living phenomena, and *biology* seemed best to him. His involvement in education reform helped spread use of the word, and by association, use of the label *natural history* decreased.

Huxley's ideas extended beyond England. Partly under his influence, when the Johns Hopkins University in Baltimore, Maryland, in 1876 established graduate training in biology, it did so by hiring both a physiologist (one of Huxley's students) and a morphologist to establish a laboratory-based graduate program. The reform of undergraduate education in the United States had the most impact on use of the term *biology* to mean a unified subject. In part to prepare students for medical studies and to support research in scientific medicine, several universities in the United States created biology departments in the 1830s that brought together specialists in the several biological sciences.

The new biology departments, staffed by young researchers trained in the latest methods, tended to emphasize the newer and exciting areas of biology, especially experimental science. Those biologists interested in traditional natural history, that is, classification and nomenclature, were concentrated in museums. This trend, noticeable by the 1890s in the United States but less so across the Atlantic, increased in the next century, and it had important implications for the reputation of natural history.

If beginning to be marginalized in universities, natural history, nonetheless, maintained a robust public profile. The fragmentation that existed under the surface scarcely could be noticed by the crowds who flocked to museums or by the thousands of readers of popular natural history. The naturalist tradition, in fact, approached its golden age at the end of the century.

The Golden Age of Natural History, 1880–1900

On Easter Sunday 1882 one of America's most famous immigrants arrived in New York City: Jumbo. The twenty-one-year-old elephant, the largest African elephant in captivity, weighed approximately six tons and stood more than eleven feet tall. He had charmed and fascinated visitors to the London Zoo for seventeen years, giving rides to thousands of children (including a young Winston Churchill), and had been doted upon by Queen Victoria and the royal family. Indeed, when the Queen discovered that Jumbo had been sold, she, along with members of various scientific societies and a sizable segment of the general population, raised an outcry. London newspapers protested the sale, and several fellows of the Zoological Society of London attempted to get a court injunction to prevent the removal of London's most famous nonhuman resident.

To add insult to the injury, the London Zoo sold Jumbo to P. T. Barnum, the "vulgar American showman" who had gone from managing a dubious private (for profit) museum in New York to promoting "P. T. Barnum's Greatest Show on Earth," a circus featuring three rings, a trapeze, and a traveling show of exotic animals. Barnum may not have had the education or expertise of the members of the Zoological Society of London, but he did know the public and its thirst for exotic and dramatic natural history. As star of the Greatest Show on Earth, Jumbo repaid Barnum many times over his $30,000 purchase and transportation cost.

But the show did not last long. After three and one half years, Jumbo met with a fatal accident when an unscheduled freight train hit him as he walked to his private rail car after a performance in St. Thomas, Ontario. Even in death, Jumbo continued to produce revenue. Taxidermists from a natural history supply house in Rochester, New York, mounted Jumbo's skeleton and hide, and Barnum displayed both to admiring crowds. Jumbo's bones finally came to rest in the American Museum of Natural History in New York, and Barnum sent the mounted skin to Tufts College for its museum. When Tufts converted the museum into a student lounge, Jumbo remained (students are said to have placed a penny on his trunk for good luck on exams), until a fire destroyed the stuffed elephant along with the building in 1975. If gone, Jumbo is not forgotten, for Tufts (now Tufts University) still features him as its mascot.

The enormous interest generated by Jumbo, as measured by gate receipts, magazine articles, children's books, and souvenirs, reflects only one facet of public interest and support of natural history. By the beginning of the twentieth century, that support reached prodigious proportions. Government-supported and privately funded universities, institutes, and research laboratories sponsored research. National, municipal, and corporate bodies built museums, zoos, and botanical gardens surpassing any previously constructed. Judging by civic and private support, most observers thought natural history had reached a golden age.

Other factors contributed to the perception. The growth of information in natural history made possible impressive publications. Catalogs, monographs, scholarly papers, and popular nature writing all attested to the robustness of the field. European colonial expansion encouraged many to see the ever-growing "conquest" of nature as an added dimension of global power. Just as exposition and museum displays of "primitive" peoples and their cultures underscored the civilizing mission of European powers, justifying intrusion into the lives of millions of Africans, Asians, and Latin Americans, so too the conquered habitats of the world on display were a reminder of Europe's responsibility to document scientifically new tracts of "virgin lands."

Museums

Nowhere did the glory of natural history appear more evident than in the construction and expansion of natural history museums in the late 1800s. Several different avenues of support came together to create these great structures. Local boosters, proponents of popular education, conservationists, imperialists, and lovers of nature found common ground in promoting the construction of "cathedrals of science," which still serve as major tourist attractions today.

The British Museum (Natural History), on the west side of London in South Kensington, first opened in 1881. Visitors thronged to see the engaging and informative public displays that occupied a sizable portion of the new building devoted to the natural history collection of the British Museum located in Bloomsbury. Such wide public use had not been traditional at the British Museum. Its original location was Montagu House, purchased by the British government to hold the extensive collections of Sir Hans Sloane, left to the nation at his death in 1753, which had men of science as curators of the

Jumbo

If there is an all-time "poster boy" for natural history, it would have to be Jumbo, the huge African elephant that thrilled London visitors at the Zoological Gardens in the 1870s and later became the leading attraction of P. T. Barnum's "Greatest Show on Earth." Everything about the animal was large—from its appetite to the size of the wagon-crate used in London to transport him from the zoo to the docks for his trip to America. It took 160 men to move his body after he was struck and killed by a freight train in Canada in 1885. Indeed, modern use of the word *jumbo* for items of astonishing size (from shrimp to jets) stems from the elephant's impact on the popular imagination.

Jumbo's celebrity reflected public interest in exotic animals and the clever marketing that capitalized on it. Showmen such as P. T. Barnum drew on the same fascination that attracted thousands of people to natural history museums, zoological gardens, and botanical gardens in the nineteenth century. It is an interest that finds contemporary expression in television nature programs and blockbuster shows in museums.

■ Courtesy of the Historical Collections, Bridgeport Public Library, Connecticut.

natural history collection. These curators possessed little interest in displaying the museum's holdings to the masses. Nor did the government initially show any effort to make the holdings more accessible.

When the British Museum outgrew Montagu House, it moved into a beautiful new building in 1852 (right next to Montagu House), but by then only half of the natural history collection fit in the space allotted; the building also housed antiquities, manuscripts, and books. Part of the collections could be displayed and consulted by scholars, naturalists, artists, teachers, and a limited number of the general public. The growth of the collections, however, soon meant that the building became hopelessly crowded. Richard Owen, superintendent of the natural history department as of 1856, served as the driving force to create yet another national museum of natural history. Owen succeeded in constructing an enormous structure in South Kensington. The imposing building with impressive research facilities and extensive public displays attracted a steady stream of visitors.

Large-scale public access to natural history museums reflected a new function for the collections. Most natural history museums at the time primarily served research needs, and the growing specialization had reinforced a professional orientation among curators, who considered the holdings for the exclusive use of serious naturalists. Natural history displays, however, had long held a fascination for the broader, general public. As earlier described, the natural history cabinets of the eighteenth century belonged to the world of fashion, and their owners assembled them for display. Enterprising individuals throughout Europe and the United States recognized the potentially lucrative value of proprietary (private) museums and natural history exhibits. Early in the nineteenth century, a ninety-five-foot whale skeleton attracted crowds in London. The curious paid to stand on a platform constructed inside the whale's rib cage and on which a small orchestra played music. Contemporaries considered it one of the highlights of the social season.

Although popular displays of nature's oddities drew crowds, their scientific worth is questionable. In 1822, three hundred to four hundred people a day each paid a shilling to see an allegedly mummified mermaid at the Turf coffeehouse on St. James Street in London. A rival merman soon appeared on display in the Strand, another fashionable street in London. P. T. Barnum, of later circus fame, sensed the public's thirst for sensational display. In 1841, this entrepreneur purchased (on credit) a private museum in New York. He soon added the contents of Charles Willson Peale's museum and continued to expand the holdings. Barnum converted a small lecture room into a theater capable of holding three thousand people and staged shows such as "Uncle Tom's Cabin." His showmanship and flair for advertising helped him attract large crowds. For many years (until it burned down in 1868) "Barnum's Great American Museum" successfully attracted the general public. Although Barnum's collection contained much scientifically valuable material, he admitted in his autobiography that the one specimen he used extensively throughout the United States to advertise his museum was none other than the mermaid from the Turf coffeehouse.

Given the association of public taste in popular natural history with the sensational, one can understand why some trustees of the British Museum had reservations about the idea that a museum should have a large public dimension in addition to its research function. The prevailing opinion among museum administrators, however, was that sensational and superficially entertaining aspects of proprietary natural history exhibits should not be confused with public instruction. Trustees of the new major museums held strong commitments to using the collections for educational purposes. At its founding in 1868, the American Museum of Natural History in New York stressed as one of its principle goals the education of the public: thousands of schoolchildren per year have visited it every year since.

The British Museum (Natural History) may have been the most famous in its day, but it represents only one of many expanded, reorganized, popular natural history museums. The Paris Muséum moved its zoological collections to a new stately structure in 1889, the same year the luxurious Imperial Natural History Museum in Vienna opened. The American Museum of Natural History in New York and the National Museum of Natural History, part of the Smithsonian Institution, in Washington, D.C., pioneered the portrayal of animal groups in environments. This practice had been developed by the commercial taxidermist Henry Ward of Rochester, New York (he was the one who stuffed Jumbo's hide). All these major museums accepted the philosophy that collections should serve the dual functions of research and public instruction.

By 1900 Germany had 150 natural history museums; Great Britain, 250; France, 300, and the United States, 250. The proliferation of natural history museums extended beyond Europe and the United States: Cape Town, Bombay, Calcutta, Montreal, Melbourne, São Paulo, and Buenos Aires opened museums that built on local collections and purchased specimens from the great commercial houses in Europe.

For the public, their enthusiasm over the museum displays often overshadowed any awareness of the considerable scientific research conducted behind closed doors. Large museums had specialized departments devoted to the study of their collections. The institutions also published memoirs, bulletins, and catalogs, as well as manuals, handbooks, and magazines. The British Museum (Natural History) pamphlet on bedbugs went through eight editions (the last one was in 1973; it still recommended DDT, since banned, for their control), and the *American Museum Journal,* started in 1900 at the American Museum of Natural History in New York (renamed *Natural History* in 1919), remains a popular magazine.

Museums extended their activities far beyond the confines of the exhibition rooms and research offices. They sponsored significant collecting explorations and collaborated with government and private expeditions that expanded the collections in scientifically interesting areas. The Paris Muséum had sent naturalists out collecting in the early 1800s, and all through the century museums sought funds to mount expeditions or competed to acquire the collections brought back by private explorers and from government-sponsored voyages. Charles Willson Peale had used many of the Lewis and Clark materials in his famous Philadelphia museum. The ambitious Wilkes Expedition of 1838–42, which returned with more than fifty thousand plant specimens and four thousand animal specimens from the Pacific, the Antarctic, and the Oregon Territory, formed the basis for the first American national museum. From the 1850s onward, all government surveys collected specimens for the national collection, which by this time had become the National Museum of Natural History at the Smithsonian Institution.

The British undertook the most famous expedition of the late nineteenth century with the H.M.S. *Challenger*, which set out shortly before Christmas in 1872 on a three-and-one-half-year voyage to explore and investigate all aspects of the deep sea. It traveled a total of nearly seventy thousand miles and brought back thirteen thousand kinds of plants and animals (along with water samples, ocean-bottom deposits, etc.). Although the *Challenger* expedition attained fame for establishing a framework for the new science of oceanography, it also contributed to the natural history of oceans. A substantial percentage of the lengthy *Report on the Scientific Results of the Voyage of H.M.S. "Challenger,"* published in fifty volumes between 1880 and 1895, is zoological studies based on material dredged or collected at the ocean's surface, and the report represents the collective research of dozens of naturalists. The amount of new biological material attracted the attention of leading naturalists worldwide. The German scientist Ernst Haeckel, for example, published a three-volume study of the beautiful microscopic marine protozoans (radiolarians) collected on the voyage. Staffs of the specialized departments at large museums such as the British Museum (Natural History) undertook the extensive taxonomic work involved in identifying new species and in considering their implications for systematics.

New techniques and technologies enhanced what traditional collecting trips could accomplish. Photography, film, and sound recordings in the early 1900s created possibilities for the study of behavior as well as for obtaining knowledge of habitats without the thoughtless destruction wrought by earlier collectors. Photography provided a method of "capture" that preserved the environment without sacrificing "sportsmanship." Carl Akeley, a noted taxidermist of the American Museum of Natural History, pioneered the use of cam-

British Museum (Natural History)

Along with Paris's national museum of natural history, the British Museum (Natural History) has been central in the history of natural history. The size of its collections, the attraction of its public displays, and its role in expeditions all contribute to its reputation as one of the greatest natural history museums. Yet a decade ago the British government considered closing the collections because of the expense of maintaining them.

The British Museum (Natural History) did not stand alone in being threatened. Natural history museums around the world agonized over their fates as governments and governing boards pondered the costs involved in operating such institutions. Among the chief public attractions of the late nineteenth century, they appeared old-fashioned to the modern

consumer of culture. The efforts of imaginative curators and museum boards, however, have given new life to these great monuments to natural history. Imaginative displays connect the collections to recent science, and the collections have taken on value as repositories of knowledge about biodiversity. The museums are still packed with crowds, and the sounds of children gasping at dinosaurs still echo in their halls.

■ *Exterior,* author's photograph; *interior,* The Natural History Museum, London (photo: W. Kimpton, 1906).

eras in natural history during the 1920s. His work popularized natural history (albeit in a rather anthropomorphic manner, which projected human motivations onto animals) and opened new avenues of research in a range of subjects, from animal behavior to stress physiology.

Zoological and Botanical Gardens

The rise of celebrated zoological and botanical gardens paralleled the development of museums. As with natural history collections, these institutions existed for centuries but fully came into their own only in the nineteenth century. Among the zoological and botanical gardens that came into prominence in the nineteenth century, the London Zoological Society's zoo had notable success in reaching a wide audience.

Sir Stamford Raffles used his social and political influence to help establish the Zoological Society. After a brilliant colonial career, first in the service of the East India Company and later as governor of Benkulen (Sumatra), Raffles returned to England in 1824. During his tenure in the East Indies he encouraged naturalists to collect and expand the knowledge of the flora and fauna of the region. His voyage home turned out to be ill fated: shortly after his ship took sail it caught on fire. Although he and his family survived, his enormous natural history collection, notes, illustrations, and living specimens did not.

Raffles, like other naturalists who experienced similar catastrophes, remained undeterred, and after his return to England he continued his efforts to promote natural history. His experiences predisposed him to stress the value of exotic specimens for a national collection, and, more important, for giving an imperial cast to a zoological garden's plans. The Zoological Society's collection, therefore, reflected the fauna of Britain's empire. Raffles and several founding members of the society, in fact, viewed the institution as significant for the import of exotic game animals (primarily fowl) which might be domesticated and used to stock the grounds of aristocratic estates.

The Zoological Society in time, however, moved in another direction: toward a popular, public zoological garden. The popularity of the Zoological Gardens in London's Regent's Park demonstrated the degree to which natural history appealed to the public. In its first year, 1828, the zoo counted 130,000 visitors; this number grew to a quarter of a million over the following decade. At first the directors envisioned the London Zoo in elitist terms. They operated it for the benefit of those able to appreciate the scientific value of the specimens, and they limited admittance to members of the society, or to the public who (upon payment) could obtain invitations from a member.

David Mitchell, who took over in 1847, actively sought to increase attendance and succeeded by focusing attention on single, dramatic acquisitions, which he advertised and promoted. In 1850, for instance, the zoo acquired a hippopotamus that drew extensive crowds and captured the imagination of the press. Mitchell also convinced the council of the Zoological Society to permit anyone willing to pay the entrance fee to visit the zoo, and this more liberal regulation considerably broadened the audience. By the 1880s the number of visitors reached more than six hundred thousand annually. Although the administration of the Zoological Gardens occasionally displeased the public—as when it permitted P. T. Barnum to buy Jumbo—its overall success helped stimulate the development of other zoological gardens and animal displays.

Across the English Channel a successful society emerged that more fully embodied Raffles's colonial goals: the Animal Acclimatization Society. Founded in 1854 to promote the introduction, acclimatization, and domestication of useful or ornamental exotic animals, the society grew rapidly. The creation in 1860 of a wonderful new zoo in Paris, the Jardin zoologique d'acclimatation represents the society's most famous venture. Like the London zoo, it enjoyed popular success and continues to this day as an important tourist attraction.

Partly because it stressed domesticated exotic species the Animal Acclimatization Zoo permitted visitors to interact with the animals. The zoo had rides, "hands-on" instruction for schoolchildren, and even the opportunity to buy specimens (alive or cooked!). The administration constructed innovative displays: one of the first large aquariums on the continent attracted enormous crowds. Although the society and its zoo failed in its original research goal—to introduce new agricultural animals to France—it nonetheless continued to display (and sell) exotic specimens and to popularize natural history.

These zoos of London and Paris had their origins partly in the attempt to domesticate exotic animals. In contrast the New York Zoological Society, which opened its park in 1899, grew out of an interest in exhibiting wild animals in natural surroundings and in helping to preserve big-game animals from extinction. Its first director, William Temple Hornaday, had been the chief taxidermist of the National Museum of Natural History. Hornaday established his reputation in Rochester, New York, in the natural history supply house that later prepared Jumbo's hide after the elephant's untimely death. Although Hornaday's early career had been spent killing animals for scientific study and display, he emerged in the late nineteenth century as one of the most outspoken advocates for the preservation of wildlife. The sentiment represented a new attitude that came to dominate natural history in the twentieth century. As director of the enormous Bronx Zoo (the popular name of the New York Zoological Society Park), Hornaday used its resources to pioneer the preservation of endangered species. His efforts to restore the American bison succeeded in keeping this symbol of the American plains alive for future generations.

Zoological gardens, like natural history museums, became a standard attraction in all major cities, and even many smaller communities built scaled-down versions. Botanical gardens showed an equal popularity, and their development paralleled those of museums and zoos. When the political leaders of the French Revolution transformed the Royal Garden in Paris into a national museum of natural history they included an extensive public botanical garden. The general public referred to the complex as the Jardin des plantes (the "botanical garden"). The Jardin des plantes housed a small zoo (formerly the royal menagerie at Versailles). This combination of zoological and botanical garden was unusual.

In London, the Kew Gardens stood far to the east of Regent's Park. Like the Jardin des plantes and other botanical gardens it had research and public functions, and like other botanical and zoological gardens it expanded significantly in the nineteenth century. Originally Kew consisted of some royal properties. During the mid-eighteenth century, owing to the efforts of the president of the Royal Society, Joseph Banks, it became a serious botanical garden. In 1841 Kew became the national botanical garden (the Royal Botanic Gardens) and began a period of impressive growth and importance, first under the direction of William Hooker and then of his son, Joseph Dalton Hooker.

Wonderful structures, such as the Palm House and Pagoda, and "plant wonders," like the giant waterlily (*Victoria amazonica*), attracted enormous crowds that rivaled those at the Zoological Society's zoological garden. Kew, however, served an important function as an institution of British imperialism. Its Museum of Economic Botany displayed the products of empire, and its greenhouses and staff belonged to a global agricultural network that orchestrated the transfer of economically important plants from different parts of the world, such as cinchona trees (important for quinine, used to treat malaria) from South America to India and rubber plants, also from South America, to various British colonies (India, Ceylon, Malaya, etc.).

The Royal Botanic Gardens and the Jardin des plantes reflect the general popularity of botanical gardens. By the 1890s more than two hundred botanical gardens existed worldwide. They attracted the support of nobility, civic groups, universities, and wealthy philanthropists of the industrial world: J. Pierpont Morgan, Andrew Carnegie, Cornelius Vanderbilt, and John D. Rockefeller stand as the first four names on the list of patrons establishing the New

York Botanical Garden (1891). Like other institutions dedicated to natural history, botanical gardens could be found in most major metropolitan areas at the turn of the century, from St. Louis to St. Petersburg.

Twilight of Natural History?

The high profile of natural history at the end of the nineteenth century, as witnessed by its institutions, reflected wide public support for the field of study. Growing popular interest in nature, and concern over the depletion of wildlife and habitats, added to the importance of natural history. In an increasingly urban and industrial world, the peace and beauty of the countryside as well as the drama of the wilderness exerted a tremendous attraction.

The proliferation of nature writing responded to an expanding and avid reading audience. Writers such as John Burroughs supplied precise descriptive essays that conveyed a deep emotional response to nature. Such works replaced the less critical popular nature writing of the nineteenth century which often contained totally fabricated tales of animal behavior (and morals).

The boom in bird watching by the late 1800s also reflected the growth of interest in nature. The Audubon Society, whose first local chapter started at Smith College in 1886, dedicated much of its efforts to protecting birds. Audubon did not have much of a role in conservation (in fact, his notebooks record what would appear today as striking insensitivity to the numbers of birds he shot), but his romantic depiction of avian life encouraged interest in birds.

The protectionist sentiment that emerged toward the end of the nineteenth century in America and beyond represented a major shift in attitude. Although even earlier naturalists had at times lamented the loss of wildlife, the efficiency of modern firearms and the wholly new scale of economic exploitation resulting from the global expansion of industrialization posed a threat that previous generations had not witnessed. The "feather trade" for the millinery, or hat making, business presented the most glaring problem. Ladies' journals in the late nineteenth century, such as *Vogue* and *Harper's Bazar,* stimulated interest in fashion, and in particular featured bird plumage for eye-catching effect. The United States imported six million dollars' worth of ostrich feathers in 1913 alone, and William Hornaday of the New York Zoological Society testified before Congress that year that he believed sixty-one types of birds were threatened with extermination by the trade in feathers for fashion. Concern over the potential loss of bird species fueled a new protectionist mentality.

Various chapters of the Audubon Society engaged support for their protectionist agenda by encouraging bird watching. Field guides for bird identifi-

Mealtime at the Zoo

Getting animals to eat in captivity presented some formidable challenges to early zookeepers. As this turn-of-the-century photograph attests, the well-known curator of reptiles at the Bronx Zoo, Raymond Lee Ditmars, resorted to the force feeding of large snakes.

■ Courtesy of the Wildlife Conservation Society, headquartered at the Bronx Zoo.

cation, the invention of prism binoculars, and bird photography greatly contributed to make bird watching easier. Annual Christmas bird counts by Audubon Society members not only supplied valuable data on the avifauna of the United States but also called attention to rare species.

The popularity of field activities and the expansion of nature writing suggest the importance of natural history at the turn of the century. The reality, as suggested by the previous chapter, may have been more ambiguous. As natural history achieved a popular triumph it experienced a scholarly decline. Universities and research institutes increasingly abandoned traditional activities of naming and classifying in favor of investigating issues in heredity and development using experimental methods. Considerable friction developed between the "amateur" bird watchers and the "professional" ornithologists, as well as between the "botanizers" and the "botanists."

But if large parts of the academic world considered natural history old-fashioned, the discipline still had its champions. The public's support of natural history museums and publications provided continued funding for research. Primary and secondary schools found "nature study" more appropriate and engaging (and less expensive) than laboratory courses. Government agencies had practical reasons for maintaining the research funding: farmers needed to know what "bugs" ate their corn; development aid workers needed to know the most productive varieties of rice, and health workers needed to identify the different mosquitoes that infested regions of the globe. Naturalists and natural history collections were still necessary.

Was the end of the nineteenth century a twilight period for natural history? For example, interest in systematics has declined through most of the twentieth century. But as we have seen, natural history is more than just naming and classifying. Since the days of Linnaeus and Buffon, naturalists have sought an understanding of the order in nature. Many early-twentieth-century biologists believed that the experimental study of living organisms would yield a new theory of biology, unifying all separate knowledge of life. Some dreamed of reducing all biological knowledge to chemistry and physics and demonstrating that all biological regularities resulted from the fundamental properties of matter; others considered various higher-level "organismic" systems that might synthesize biology. The naturalists' goal of searching for the order in nature, however, had not been abandoned.

Naturalists, it turned out, proved to be more vital to this larger enterprise than many biologists in the early 1900s would have predicted. Far from entering a period of decline, natural history provided the basis for the principle theoretical synthesis of the life sciences in the twentieth century and for a new set of practical concerns.

8 New Synthesis

The Modern Theory of Evolution, 1900–1950

The Water Babies, published in 1863, delighted several generations of English-speaking children. Charles Kingsley, a close friend of Thomas Henry Huxley and a prominent Victorian clergyman, wrote the book as a fantasy to demonstrate a parallel between the natural and spiritual worlds. Fairies transformed children who had been mistreated or had died of preventable diseases into "water babies" (four-inch, humanlike creatures with external gills). Through moral evolution, they were able to grow up and become adult humans. Kingsley's charming story plays with ideas about what we can see and know and challenges children to question whether things they cannot see (water babies) actually exist.

In 1892 at the age of nearly five, Julian Huxley read Kingsley's book. Intrigued by an illustration of his grandfather, Thomas Henry Huxley, and Richard Owen examining a bottle that supposedly contained a water baby, he wrote to his grandfather to ask whether he had seen a water baby and whether he (Julian) might someday see one. The elderly Huxley (he died three years later) waffled on the issue of what he had actually seen but told his grandson: "There are some people who see a great deal and some who see very little in the same things. When you grow up I dare say you will be one of the great-deal seers and see things more wonderful than Water Babies where other folks can see nothing."*

Julian Huxley did just that. Admirers thought that he had vast synthetic—at times visionary—ability. His critics thought he saw more than actually existed. From an early age he found living things fascinating. In his autobiography he wrote that his first memory (which dated from age four) was of a toad that jumped out of a hawthorn hedge. Whether or not that event led to his career as a scientific naturalist, as Huxley claimed, he did develop a strong interest in natural history, and in the life sciences more generally. Bird watching at Eton led to a serious interest in ornithology, especially in the behavior of birds,

*Julian Huxley, *Memories* (London: George Allen and Unwin, 1970), 24–25.

and as a student at Oxford he explored the evolutionary foundations of bird behavior.

Huxley did not limit himself to the traditional subjects of natural history. A year's fellowship in 1909 at the Zoological Station in Naples allowed him to experience the excitement (and often the frustrations) of experimental biology. He experimented on sponges by separating the organisms into individual cells and then following how they regrouped and developed. Although he published the studies, he felt dissatisfied with his abilities in physiological research—but not enough to abandon them completely. He went on to do serious investigations of growth and development.

His first love, natural history, was the perspective that informed all his work, but it was deeply infused with knowledge from all branches of biology. This can be seen in the volume of essays he edited in 1937, *New Systematics*. The book reflected the work of the Association for the Study of Systematics in Relation to General Biology, one of several groups in different countries which focused on revitalizing taxonomy. During the third and fourth decades of the twentieth century, naturalists, particularly in the United States, Great Britain, and the Soviet Union, hoped that the pioneering work being done in cytology, ecology, and genetics might provide the basis for an "experimental taxonomy." In spite of initial enthusiasm, the new systematics proved to be problematic. The fields capable of illuminating taxonomy were themselves experiencing considerable flux, and the individuals involved in classification rarely had time to bring themselves fully up-to-date on the latest developments. Cytological or ecological knowledge applied to taxonomic questions allowed a broad range of interpretation, and by midcentury most naturalists realized it was unlikely that the new systematics would eliminate the conundrums of taxonomy. Experimental taxonomy, however, did provide many new techniques and tools for taxonomic work.

Theodosius Dobzhansky, a Ukrainian naturalist who in 1927, early in his career, came to the United States to work with the geneticist Thomas Hunt Morgan, made striking use of the new tools. Dobzhansky clarified the taxonomic relationship of what had been considered two major geographical races of a single species of fly (*Drosophila pseudoobscura*). By using staining techniques developed to study cells, he and his group identified structural rearrangements on the chromosomes (easily seen in the gigantic chromosomes of the fly's salivary glands) of the two races, and these rearrangements could be used to help differentiate the races. Although known to produce sterile offspring when interbred, the two races appeared morphologically identical and overlapped geographically. They differed, however, in a variety of physiological functions. Ultimately, it was the genetic and chromosomal differences that proved central in convincing Dobzhansky and others that the two races were actually separate species.

Not many species had chromosomes as easily seen as those of *Drosophila*, so Dobzhansky's techniques would not be generally applicable. But that was not the point. Taxonomists came to realize that the new biological specialties had tools and techniques that could be incorporated into an understanding of biological relationships. Because many of these relationships were evolutionary ones, the new systematics helped bring an even closer tie between taxonomy and research on evolution. It was no accident, then, that many researchers associated with the new systematics were important in the reformulation of the modern theory of evolution.

The Modern Synthesis

Life scientists in the late nineteenth century had come to general agreement that living organisms evolved over time, but by the dawn of the twentieth century there was little consensus on how that evolution occurred. Darwin's emphasis on natural selection seemed overly simplistic to many. Scientists proposed various alternatives that stressed either large-scale, sudden changes or general, progressive "tendencies."

The rediscovery of Mendel's Laws in 1900 would later be of central importance, but it initially distanced many of Darwin's supporters from the research done in heredity and from what ultimately came to be the specialty of *genetics*. Central to Mendelian genetics is the notion that hereditary traits are discrete units that come from parents and can recombine in the fertilized eggs that become offspring. Understanding how traits recombine appeared to many researchers the key to evolution. The leading figures in genetics criticized the "outdated" selectionist (i.e., Darwinian) approach, but that did not discourage a central core of naturalists. In particular, several of the leading figures of the new systematics movement, notably Julian Huxley and Ernst Mayr, remained fundamentally Darwinian in their orientation and brought that perspective to the entire body of their work. It was, however, the pioneering research done in population genetics by Soviet scientists, many of whom had naturalist backgrounds, which proved decisive in uniting evolution and modern genetics.

Theodosius Dobzhansky began his career in Kiev in 1918 studying the systematics of ladybird beetles. His move to the genetics institute in Leningrad shifted him to a different line of research, but his natural history background remained an important influence. Dobzhansky studied the genetics of popu-

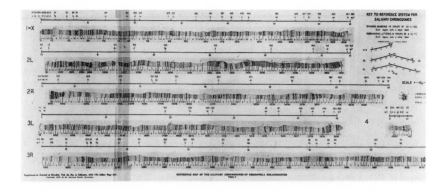

Choice of Research Animals

Drosophila, the common fruit fly, played a special role in the Modern Synthesis. Its study helped establish modern genetics, and research on its distribution contributed vital knowledge in population genetics. Why fruit flies? In part, they are easy and inexpensive to maintain and study. They require two weeks to complete a generation; a single pair can produce hundreds of offspring. The laws governing the inheritance of their physical traits are simple. Equally important, they possess giant salivary gland chromosomes that can be easily stained to study their structure.

The *chromosome map* shown here demonstrates what geneticists could do in the 1930s. For example, Theodosius Dobzhansky made use of such maps to conduct research on *Drosophila* and to show important relationships among different populations.

■ From *Journal of Heredity* 26, no. 2 (1936); courtesy of Oxford University Press.

lations found in nature, and his classic *Genetics and the Origin of Species* (1937), which he wrote after moving to the United States, argued for a return to emphasizing Darwin's concept of natural selection in evolution.

Unlike the history of other great theoretical shifts, the emergence of the Modern Synthesis had several classic texts rather than one. The difference reflects the much larger nature of modern scientific communities, which are communal enterprises, in contrast to earlier times, when the scientific investigator was more isolated. Dobzhansky's book, the first of the seminal texts that defined the Modern Synthesis, sketched the outlines of the modern theory of evolution. Other texts quickly followed. Ernst Mayr (curator of birds at the American Museum of Natural History in New York and previously leader of three expeditions to New Guinea and the Solomon Islands from the Zoological Museum at the University of Berlin) wrote *Systematics and the Origin of Species* (1942) from a naturalist's perspective and stressed the importance of geographical variation. Julian Huxley's *Evolution: The Modern Synthesis,* published the same year, supplied the most popular name used for the modern theory of evolution. Like Mayr, he took his starting point from the natural history tradition. George Gaylord Simpson's study *Tempo and Mode in Evolution* (1944) used the fossil record to establish the compatibility of the Modern Synthesis with paleontology. Later, G. Ledyard Stebbins's *Variation and Evolution in Plants* (1950) brought plants fully into the story.

Several different lines of research converged in the Modern Synthesis. The naturalist tradition played a critical role by supplying a conception of species which expanded Darwin's insight and made clear a distinction central to Darwinian evolution. *Species* before Darwin had traditionally been defined in terms of physical characteristics. Occasionally naturalists used other criteria, for example, Buffon recommended successful interbreeding as a test of relatedness. Given the practice in natural history of relying on preserved museum specimens for classification, the use of anatomical features, usually external, had predominated. A *type specimen* referred to an individual specimen, usually in a known collection, that served as a physical model to define the species. *Type species,* similarly, referred to individual species that embodied the shared characters of a genus. Darwin's shift to a dynamic view of nature undermined the meaning of types because he considered species as groups of individuals; the composition of the group—due to variation and selection—changing in time. A species, therefore, needed to be thought of as a population, or more accurately, a set of populations. The ornithologists whom Mayr studied under in Germany had employed a version of this conception of species. Mayr, who had field experience as well as museum training, emphasized the need to reconceptualize older concepts of species, or what he called the *typological* concept of species which stressed a static plan, and to draw out the populational thinking implied by Darwin in light of new knowledge.

An understanding that species had isolating mechanisms that prevented successful cross-species reproduction lay at the heart of the new concept. Mayr's *Systematics and the Origin of Species* proposed that a new species develops if a population becomes geographically isolated from its parental species population and its members develop characteristics that make it impossible to successfully reproduce with individuals from the parent population should contact between the populations be renewed. The perspective reflected Mayr's geographical research on birds. Mayr proposed a new definition of species, which he called the *biological species concept.* Expressed in the simplest terms, biological species are "groups of actually or potentially interbreeding natural

populations, which are reproductively isolated from other such groups."* Even
though Mayr's characterization underestimated other recognized methods of
speciation that exist in plants and various lower organisms, it quickly caught
on because it captured both the natural history background and the genetic
underpinning of evolutionary thought.

Critical as Mayr's work was for the Modern Synthesis, historians more often cite Dobzhansky's *Genetics and the Origin of Species* as being of central importance, in part because it came out first. More important, the Modern Synthesis emphasized the genetics of population change. The collaboration of
mathematical modelers, laboratory scientists, and field biologists brought the
study of the genetics of natural populations, the genetics of specific traits in
individuals, and the theoretical models of population dynamics together into
a new discipline of *population genetics*. Pioneers in this field sought to demonstrate how natural selection operating on small characteristics could result in
larger, stable evolutionary change.

Dobzhansky, who in his early research had been interested in the amount
of variation in natural populations, took techniques from recent work in genetics and applied them to the study of natural populations of *Drosophila* to
demonstrate the genetic variation within and between species. Dobzhansky
also collaborated with the American mathematical geneticist Sewell Wright,
who had done experimental and theoretical work in both a university setting
and in the Animal Husbandry Division of the U.S. Department of Agriculture. Wright stressed the interaction of gene systems and, in what he termed
random genetic drift, the chance fluctuation of gene frequencies in small populations. The rigor of the mathematical models and the tie to experimental results lent greater scientific respectability to the theory of evolution. Given that
scientists had criticized earlier theories of evolution as too speculative, the field
of genetics proved valuable by providing an impressive and rigorous material
basis for a new theory of Darwinian evolution.

Unifying Strength of the Modern Synthesis

In the wake of the publicity surrounding the dramatic research being conducted today in molecular biology and in molecular genetics which has resulted in genetically engineered crops and the identification of the genes responsible for specific diseases, one could easily lose sight of the Modern
Synthesis's tremendous intellectual importance. The modern theory of evolu-

---

*Ernst Mayr, *Systematics and the Origin of Species* (1942; New York: Dover, 1964), 120.

tion represents a unifying project more ambitious than Darwin's earlier synthesis. Late-nineteenth-century life scientists such as Haeckel had argued for a synthesis based on evolutionary principles, but the lack of a sufficiently coherent and accepted theory of evolution impeded the development of a unified biology. The architects of the Modern Synthesis deliberately set out in the 1930s and 1940s to construct a theory of evolution that would make sense of what they already knew of the living world. Their success rested on acceptance of a formulation of natural selection with population genetics as the driving force of evolution. Random mutation, recombination, and selection drove the evolutionary process, and everything in biology could be understood as a consequence. Dobzhansky's famous quip that nothing in biology made sense except in the light of evolution best sums up the enormous generalizing power the Modern Synthesis claimed for itself.

The early proponents of the Modern Synthesis also saw evolution in broad philosophic terms and thought that as a worldview it had social significance. Julian Huxley, perhaps more than any of the other creators of the modern theory of evolution, elaborated on the wider implications of evolution.

Huxley claimed in *Evolution: The Modern Synthesis* that biology had embarked on a phase of unification following a period in which emerging disciplines had developed in isolation. He held that in the study of evolution varied facets came together: ecology, genetics, paleontology, distribution, embryology, systematics, comparative physiology, and comparative anatomy found common ground in the study of evolution. The unification of biology held even more promises, however. Huxley claimed that the study of evolution revealed biological progress over time. But what of the future? Although he argued that the study of evolution did not reveal purpose in nature, it did show direction: toward an increase in control, independence, and knowledge, and toward the means of coordinating knowledge. Human evolution reflected those advances and stood as the latest stage of an ancient story.

Typical of a large segment of Western intellectuals at the time, Huxley sought a replacement for the crumbling Victorian worldview that had been undermined by the First World War. For him, natural science promised to be this replacement, and evolution, the science of origins, appeared the most likely to chart human destiny. Huxley argued that humankind had reached a stage of evolution where it created values and that future human evolution would entail the increase and advancement of aesthetic, intellectual, and spiritual experience. In numerous essays Huxley elaborated on a scientific humanism that he characterized as basically democratic and progressive and that stressed education and humanitarian concerns based on scientific knowledge.

Naturalists in the Field

Field biology has improved considerably since the heroic days when naturalists traveled off to distant lands without the protection of vaccinations or the lifelines of radios and cell phones. Many early naturalists of the late eighteenth and early nineteenth centuries did not survive their journeys. Collecting in the field is still hazardous, and naturalists must contend with medical, political, climatic, and logistical problems. But the attraction remains.

Ernst Mayr collected birds in Dutch New Guinea, Papua New Guinea, and the Solomon Islands between 1928 and 1930. His observations on the distributions of birds later proved critical in his evolutionary thinking. This photograph of Mayr

and his Malay assistant was taken in Dutch New Guinea in June 1928.

■ Courtesy of Ernst Mayr.

Other architects of the Modern Synthesis explored the human implications of evolution with fairly similar conclusions, resulting in a liberal humanism that stressed the value of science and the mechanical and material basis of life. Like earlier naturalists who framed their vision of nature in terms of contemporary ideas, the modern evolutionists crafted a consistent and highly popular image that suited the political and intellectual concepts of the day. Just as Linnaeus located his taxonomic contributions in the context of a northern Christianity, and Buffon constructed a tableau that he offered as a cornerstone to the French Enlightenment's efforts to reform European thought, so, too, did Huxley and his associates envision their brilliant biological synthesis as part of a larger cultural construction. The strength of the biology, they held, lent support to their shared social and philosophical convictions. Although the architects of the Modern Synthesis regarded evolution as part of a wider worldview, we should not confuse, however, the vast unifying power of the theory with an acceptance of its alleged philosophical or social implications.

The modern theory of evolution unifies the life sciences in a way that early naturalists had hoped their science would. The naturalist tradition sought to describe the living world and to discern its order. Darwin's theory of evolution grew out of his attempt to understand how different species came into existence and their relationships to one another. In so doing, he fashioned a theory that explained other regularities and made sense of why living organisms functioned as they did. Physiologists uncovered how organisms functioned and showed the amazing complexity and integration of systems. Evolution, by relating function to adaptation, showed how and why functions came into being. The Modern Synthesis used the fruits of decades of experimental biology to help elucidate the material basis of evolution and to bring a deeper understanding of its mechanisms. By its profoundly historical nature, the theory continues to demonstrate the importance of the naturalist tradition. Only by studying what has actually come into existence; only by uncovering the extant fossil record to determine which phylogenies have developed in the course of time; and only by documenting the distributions of contemporary and extinct species can we know life on Earth.

The life sciences have uncovered many general laws, such as Mendel's Laws of inheritance and the genetic code. But the theory of evolution goes beyond significant and interesting regularities common to living organisms by viewing all biological knowledge as the result of a long historical process. The Modern Synthesis, in an important way, fulfills the goal of earlier naturalists to describe and comprehend the order in nature. It is not what Linnaeus or Buffon would have expected—and, like all of our central unifying scientific theories, it remains to be seen how future generations will alter or regard it—but it currently stands as a major milestone in the naturalist tradition.

E. O. Wilson, 1950–1994

Edward O. Wilson begins his autobiography, *Naturalist*, with two descriptions of his earliest memories of the power and mystery of nature: encountering a large jellyfish in the shallow water off Paradise Beach at the age of seven, and observing a gigantic stingray from a dock in the same part of Florida's Perdido Bay during that summer of 1936. The little boy's sense of the wonders of nature never seems to have left Wilson, who grew up in Alabama and spent much of his childhood catching snakes and classifying ants. He pursued his interests at the University of Alabama at a critical time for American biology: during the spread of the new evolutionary theory, that is, the Modern Synthesis. Several of Wilson's professors had come from the centers where biologists had constructed the Modern Synthesis, for example, the American Museum of Natural History in New York, where Ernst Mayr, curator of birds, worked.

Mayr, the most energetic proponent of evolutionary thinking in the United States, stressed the importance of studying geographical distribution and called attention to the theory's new concept of species. He also understood its implications for classification: classification should reflect evolutionary relationships and embody the knowledge of evolutionary differences that have occurred over time. In practice this approach encouraged a careful recording of geographical ranges and thorough comparison of differences.

The evolutionary perspective remained central to Wilson's later work. After a brief stint at the University of Tennessee, Wilson entered Harvard University for graduate work and has spent his entire career there. Not confining his research to merely collecting and naming ants, he has been able to incorporate the methods of other branches of the biological sciences into his research, either directly or indirectly, by collaborating with other scientists. In addition to traditional classification, he has been involved in extensive field collecting, mathematical modeling, experimentation, theorizing, and philosophical reflection.

Ants were Wilson's entrée into the order of nature. The tremendous specialization of the past hundred years has been a double-edged sword. For many individuals, it has narrowed their focus of vision and concentrated their energies on a set of specialized questions that only a small number of equally trained scientists find interesting. Increasingly, broad theoretical issues have inhabited a no man's land. But the mastery of a small group of plants or animals has also been a powerful wedge for a few, highly creative biologists, who by being alert to the opportunity have been able to ask interesting questions that they can explore with the highly specialized knowledge they possess.

In this "opportunistic" exploration, as he terms it, Wilson has been one of the most successful biologists of the twentieth century. Although he devoted years of study to the traditional activity of classifying ants, early in his career he asked questions about their behavior, distribution, ecology, and evolution. Stimulated by a lecture given by the German ethologist Konrad Lorenz on animal behavior, for example, Wilson set out to examine what could be learned about the chemicals ants use in communication. Ethologists like Lorenz argued that animals responded to signals (from the environment or from other animals) that triggered fixed-action patterns. For example, if an egg rolls out of the nest of an incubating graylag goose, and if she sees the egg, a set of patterned actions will follow, with the result that she rolls the egg back into the nest. To ethologists the behavior appeared fixed and innate. Their experiments revealed that graylag geese would return to the nest any egglike object (and even some only slightly egglike, such as a metal can) they noticed nearby.

Ethologists described many such signals (called *releasers*) in mammals, birds, fish, and insects; most were visual or auditory. Wilson knew that ants and other social insects—because of the darkness of their nests and their supposed inability to hear sound through the air—had to rely on other means of communication. Scientists knew that ants transmit information by way of tapping their antennae and forelegs, and in one case there seemed to be a chemical substance used to mark trails. Wilson constructed an artificial nest for fire ants and watched their behavior. He noted how they communicated information about the site of food by means of odor trails. Next he sought the source of the chemical used in laying the trails. Through microdissection he systematically removed different internal organs from the ants and created artificial trails with extracts made from the organs. After numerous trials, he discovered the small gland which produces a substance the ants use to indicate the location of food, a substance which also acts as a strong stimulus to the fellow worker ants to search for the food. Collaborating with other scientists, Wilson learned the general characteristics of the chemical, and later research identified other chemicals involved. Wilson explored the chemicals that ants employ for other communication, and he began the study of the general properties and evolution of these chemical signals, called *pheromones*. Many animals employ pheromones, as it turns out, and Wilson's research opened up a rich field of research.

Wilson's career illustrates the broad landscape in which the naturalist can operate. Wilson contends that the naturalist occupies a privileged position: to stand back and, through the lens of a particular biological group, catch glimpses of general features of the order of nature. His opinion runs against much of the direction of contemporary research. Earlier specialization in natural history limited naturalists to studying individual biological groups (birds, fish, etc.). The trend toward increasingly specialized research has intensified owing to new experimental methods, so that today biologists focus on particular functions of groups (e.g., plant cell biology). This narrowing has occurred despite the fact that the Modern Synthesis integrated knowledge from different biological groups and from different levels of organization.

Some of these new fields have been compelling in their results. Spectacular discoveries in molecular biology and gene research have mesmerized the public and its funding agencies because of the promise of dramatic gains in medicine and agriculture and a possible revolution in the social sciences and humanities. Wilson argues, however, that many "generalizations" emerging from molecular studies of specific groups have been limited to the groups under scrutiny and are not applicable to life in general. In contrast, by examining a specific group of organisms on several levels, the naturalist can uncover generalizations that transcend levels of organization and that have relevance for broader groups of organisms. Wilson's research on ant behavior, for example, has taken him into the realms of biochemistry, ecology, and evolutionary theory and has significance for the entire animal kingdom.

Biodiversity

Wilson calls for a keener appreciation of the naturalist tradition. Natural history's concern with specific taxonomic groups (taxa) can lead to new biological insights. The careful study of taxa has other important value. Wilson has been a central figure in promoting the need to document the diversity of life on Earth. Biologists to date have named and described a total of roughly one and a half million living species. Compared to what Linnaeus and Buffon knew that sounds enormous, but it represents only a small portion of what exists. Depending upon which expert one asks, the estimates of the number of contemporary species on the planet vary between five million and thirty million. Clearly, much remains to be done.

What has given the task a new urgency has been a recognition by biologists that as the human population expands it alters the Earth's environment, thereby threatening extinction to numerous as-yet-unknown species. Even more disturbing, biologists have discovered that the rate at which species become extinct today approaches the rate seen in those periods in the fossil record that we have labeled *catastrophic extinction events*. Scientists particularly worry about the fate of the tropical rain forests, which, although they represent merely 7 percent of the planet's terrestrial surface, serve as home to more than half the living species in the world. Despite the rich diversity and density of vegetation, these rain forests are surprisingly fragile systems. Once disturbed they regenerate slowly, if at all. Large portions of the rain forests known to nineteenth-century explorers have been destroyed. Coral reefs and coastal wetlands face equal danger.

Does it matter? A large international movement of concerned individuals thinks so, and under the banner of "saving biodiversity" have sought to call attention to the problem and to influence governments and international agencies. Wilson has been their leading advocate and has succeeded in attracting international publicity on the need to fund research to document what is left of the planet's biodiversity. In 1986, the National Academy of Sciences and the Smithsonian Institution sponsored one of the first major forums to focus the public's attention on these issues. More than sixty recognized experts in biology, economics, philosophy, and politics explored the implications of the threat to biodiversity. Since then, the United Nations has hosted international meetings which have tried to balance conflicting interests. The demands of economic planners in countries with exploding population growth oppose the concerns of biologists over the effects of economic development on threatened biological resources.

Participants in international meetings, such as the United Nations Conference on Environment and Development in Rio de Janeiro (1992), have concluded that human activity causes an alarming loss of diversity, but they have not been able to agree on what steps should be taken to remedy the problem, or even on estimates of the actual loss. In these discussions one point on which there has been consensus concerns the importance of assessing the damage to biodiversity. Lack of knowledge of the organisms that inhabit the planet hinders our understanding of the actual loss. Until some baseline can be established, it will be impossible to evaluate the conflicting claims of what has happened in places like the tropical rain forests or coastal wetlands.

The great natural history museums of the world provide repositories of knowledge about life on Earth but have limited resources. Wilson and other naturalists have called for a renewed commitment to the classic goals of nat-

A

B

Young Naturalist

From an early age, E. O. Wilson was fascinated with the diversity of life. (Shown here as a thirteen year old [A], he collects insects in Alabama.) His autobiography describes his attraction to the natural world and his lifelong dedication to observing its bounty.

The twentieth century has produced many notable naturalists, but few can match the breadth of Wilson's interests and contributions. Like the notable naturalists of the eighteenth century, Linnaeus and Buffon, Wilson has sought to document and interpret the details of nature as well as make broad generalizations about it. He has wondered how humans fit into this picture, often venturing into rough terrain—as in Central America (B)—in his attempt to synthesize knowledge from a biological vantage point.

■   A, courtesy of E. O. Wilson; B, courtesy of Minden Pictures (photo: Mark W. Moffett).

ural history to provide the knowledge necessary to document biodiversity. They point out that even in an optimistic projection of what it would take to complete a global biological survey, there would be large international teams, working over a period of fifty years, at different levels or scales of time and place. First, threatened habitats containing the greatest number of endangered species would be identified and inventoried. Inadequately known ecosystems would be studied to see whether they contain threatened local habitats. Next, wider areas thought to be threatened would be identified, and research stations established to monitor environmental factors; and then the extensive task of documenting the flora and fauna would begin. Last, the results of local and broad area surveys, supported by monographs written on individual groups, could produce a useful picture of the planet's biodiversity.

The modern theory of evolution and recent concern with biodiversity demonstrates that the naturalist tradition in the twentieth century has strong continuities with the past. The traditional goals of naming, describing, and ordering have remained important. The earlier quest to complete a catalog of nature has turned out to be a much larger enterprise than imagined, but still an important one. Similarly, the general picture of nature that emerged from the Modern Synthesis provides fertile research questions and continues to develop to ever finer levels of resolution.

Natural History and Biology: Ecology

Natural history in the twentieth century has influenced areas of the life sciences usually thought to be outside or different from the naturalist tradition. The history of population ecology offers a striking example. Researchers in the middle of the twentieth century sought to move ecology from documenting case histories that examined interactions in a local setting in a largely descriptive manner to the use of experiment, mathematical modeling, and abstraction in search of universal laws. George Evelyn Hutchinson, for one, stressed the value of mathematical models in studying nature. He believed that such models could be combined with careful fieldwork and experimental design to produce a science that went beyond knowledge of specifics to uncover patterns in nature. Hutchinson's students in the 1950s helped transform the discipline of ecology. They argued for combining knowledge derived from field observations with mathematical modeling to predict currently unrecognized relationships. Observation would then determine their predictions' validity. They also believed that their research would uncover general patterns and principles.

Wilson occupies a somewhat curious role in this story. He and his collaborator, Robert MacArthur, a student of Hutchinson, wanted to change biogeography, the study of geographic distribution of organisms, from a subject limited to description of the distribution patterns to a science that proposed testable hypotheses. In a sense they sought to move away from conventionally conceived natural history. They based their theory on the observation that on an island, species diversity could be correlated with area. Ernst Mayr and other naturalists had contributed the observational basis for those correlations. Mayr called attention to records suggesting that the existence of different animal species on islands changed more rapidly than expected. He thought this change resulted from extinctions of vulnerable immigrant populations.

Looking for a more general solution, MacArthur and Wilson proposed that animal diversity resulted from a dynamic equilibrium (a changing balance) of immigrations and extinctions. Rather than identifying species, they sought to identify an underlying relationship of immigration and extinction rates. They did so using two mathematical curves, each representing one of the rates. The point of intersection of the curves predicted the number of species at the balance point, or equilibrium. MacArthur and Wilson used this simple relationship to consider questions about general characteristics of immigrant, or colonizing, species and rates of extinction. They suggested, for example, that selection in a crowded environment differed from selection in an uncrowded one and that the evolution of species would proceed according to prevailing conditions. They also claimed that the validity of their theory could be verified by observation and experiment. Their 1967 book, *The Theory of Island Biogeography*, defined a new style in ecology and greatly influenced research.

For Wilson, island biogeography grew out of his own search for generalizations. He soon turned to other subjects—animal behavior, and then biodiversity. The population ecology he helped pioneer, however, developed along mathematical lines, which, when combined with fieldwork, stressed the study of current ecological change over relatively short time-spans (compared to long-term evolutionary change). The emphasis on mathematical generalization downplayed the individual species involved as well as their histories in favor of more abstract formulas.

As the great defender of the naturalist tradition, Wilson ironically helped spawn studies that came to reject the value of natural history. Even so, the importance of this supposedly antiquated knowledge soon became evident again. Scientific concern with environmental degradation and declining biodiversity raised questions that exposed the limits of mathematical models. When ecologists compared the biodiversity of similar areas (in terms of climate, soil, etc.), they discovered differences that could be explained only by studying the particular historical backgrounds of the areas. That is, the general mathematical models failed to account for the differences, but if supplemented with descriptive and historical information about each area, the differences could be explained. Conservation biology, which attempts to protect endangered ecological systems and to restore habitats, poses practical problems that raise similar issues. In tackling problems, therefore, ecologists came to rely on the history of plants and animals along with models of their interactions. The "new" natural history in ecology blends the rigorous formulation of mathematical models and the development of testable hypotheses with the long-term historical backgrounds of particular areas.

Natural History as Worldview

Linnaeus and Buffon pursued their research in natural history within a broad eighteenth-century framework that assumed the orderliness of the world and the ability of humans to discern that order. Although they both thought of plants and animals as only part of the totality of existence, they strongly believed that the knowledge they uncovered held wider significance: the order of nature contained lessons and informed a range of human interests. Darwin's theory of evolution, likewise, inspired nineteenth- and twentieth-century thinkers to explore the wider meanings of Darwin's vision of the living world. Architects of the Modern Synthesis, such as Julian Huxley, extended Darwinian ideas in biology and speculated about the relevance of evolution for humankind.

Contemporary natural history continues to seek the deeper philosophical and social implications of biological knowledge. E. O. Wilson, who may best exemplify the twentieth-century naturalist, has argued that the detailed study of specific taxa can lead to far-reaching generalizations. He also believes that a general unification of knowledge can be obtained through an extension of biological and physical thought "across the borders" into the humanities and social sciences. Wilson unabashedly allies himself with the Enlightenment quest for a unified understanding of the physical and human world. Evolution provides the key to unlocking the mysteries of mind, society, ethics, and art, he writes; natural selection explains the facts of classification, distribution, the fossil record, embryology, and behavior. To Wilson, we have only begun to grasp the explanatory power of viewing evolution as a *process*. Together with a material understanding of the brain, an evolutionary understanding of the central problems addressed by the humanities and social sciences can resolve perennial human questions.

Wilson acknowledges the speculative nature of his vision, but he argues for its potential as a research agenda. What he has in mind is a study of the relationship of genes and culture. Or, to phrase it in a different manner, a study of how human *genetic* evolution has interacted with human *cultural* evolution. Wilson believes that the two forms of evolution have evolved in a mutually dependent manner, analogous to the interactive *coevolution* of some plants and animals, such as the adaptations of insects and the flowers they pollinate. Because the brain as an organ has evolved over millions of years in response to natural selection, Wilson notes, biologists should pay greater attention to it as a machine involved in the battle for survival.

Wilson pioneered the subject of *sociobiology*, a synthesis of animal behavior and evolutionary biology that raised a storm of protest in 1975, when he published a 700-page treatise that laid out the broad outlines of the subject. The protest focused on his last chapter, which stressed the biological bases of human behavior and even suggested that evolutionary biology had reached a stage of development at which it could address issues in ethics and other subjects formerly considered the province of the social sciences and humanities.

Drawing on research in evolutionary psychology and sociobiology, Wilson contended that "prepared learning," in which humans and animals are innately prepared to learn some behavior (or resist learning) over other behavior, serves as a link between genetics and culture. Evolutionary psychologists believe that these predispositions hold evolutionary significance. A concrete example Wilson uses is the ease with which most humans learn to fear snakes. Animals closely related to humans, such as long-tailed arboreal monkeys of Africa, instinctively emit danger calls when they see various poisonous snakes. Although children under five years old show no fear at seeing a snake, they appear increasingly uncomfortable as they mature, and it takes merely one or two negative experiences to render them permanently afraid for the rest of their life. Because many snakes are poisonous, avoiding them confers a selective advantage for survival. Humans, of course, can overcome their fear (otherwise the discipline of herpetology might not exist!). Evolutionary psychologists claim that this flexibility demonstrates the complexity of prepared learning: it is partly inherited and partly learned.

According to Wilson, prepared learning, might apply to social behavior. Human societies have existed long enough, Wilson argues, that adaptive social practices have been selected and survive in the human gene pool in the form of genes responsible for various forms of prepared learning. Universal human practices that anthropologists have documented, therefore, may be explicable in evolutionary terms. The fascination with certain images in widely disparate cultures—for example, the omnipresence of serpents in myths—may reflect deep impressions that find cultural expression in stories.

Wilson has crafted an ambitious program. At its core stand the insights of natural science, in particular the modern theory of evolution, with its links to classification and description, to genetics, and to neurobiology. Evolution makes sense of the facts of natural history and may be instrumental in illuminating central concerns of the humanities and social sciences. Wilson contends that until students of the human condition take our evolutionary and biological nature seriously, we will not advance our understanding.

Critics, as might be expected, have reacted with great indignation at the suggestion that the hallowed disciplines of philosophy, history, and literary criticism, and the worldly subjects of economics and political science should be unified with the natural sciences and thus yield pride of place to biology as the grand unifier. Wilson's position strikes many humanists as too general and too simple to ever lead to a detailed understanding or to the uncovering of specific evidence to substantiate it. How can we penetrate the minds of long-dead paleolithic peoples to study how their ideas naturally developed? Wilson's critics have also charged that he advocates a biological determinism and that his ideas carry the danger of abuse similar to earlier eugenic ideas with their overtones of racism and sexism.

Like the creators of other synthetic systems, Wilson passionately believes that he has found the key to unlock the mysteries of nature—physical, biological, and human. Similar to Enlightenment naturalists and the architects of the Modern Synthesis, he seeks a unity of knowledge and argues that science provides the unifying method. He claims his vision possesses a strongly humane dimension, one compatible with the democratic and humanistic values of our culture.

Wilson stands in a long line of naturalists who have sought an understanding of the order in nature and who have blended that knowledge with human concerns. His contributions to systematics and evolutionary theory ensure him a prominent place in the field of natural history, and his thoughts on culture contribute to an ongoing dialog in contemporary thought on the foundations of ethics, politics, and art.

Epilogue

In 1735 the young Linnaeus printed a dozen pages outlining a classification system for animals, plants, and minerals. Four years later Buffon took over management of the Royal Garden in Paris and soon began planning his encyclopedia of natural history. Each pioneer labored for more than four decades and produced volumes that together established an essential foundation for the life sciences. The writings of Linnaeus and Buffon surveyed all that late-eighteenth-century Europeans knew of the living world. The two men had different motives. Linnaeus sought to catalog God's Creation and catch glimpses of the order that underlaid His plan; Buffon aspired to provide his generation with a secular natural history on a scale not attempted since the work of the Roman naturalist Pliny and to uncover the laws of nature that formed the patterns in his broad tableau. Buffon's *Histoire naturelle* reached a level of general popularity so high that anthologies of French literature still include it. Linnaeus had an international following, and if his publications lacked the stylistic elegance of his French rival, popularizations of his work nevertheless attracted an enormous number of enthusiastic readers.

In the two and one half centuries since then, naturalists have made impressive contributions. They have documented some important groups, such as birds and mammals, almost completely. The theory of evolution is widely believed to unify and explain the phenomena of life. Ecology and environmental biology study plant and animal interactions and the effect of humans on them. Today we speak of a human "stewardship" for the life on Earth which reflects a sense of our growing ability to influence the course of nature and our enhanced sense of responsibility for the natural world.

Just as Linnaeus and Buffon realized that their efforts marked only the beginning of a quest to examine and understand nature, so today, with our vastly increased ability to examine the world around us, naturalists marvel at the world yet to be discovered. The threat to biodiversity has brought this point forcefully home. The microbial world remains as foreign to us as the New World animals and plants were to the naturalists of the early eighteenth century. Even for those animals, plants, and microbes for which we do have names, we know next to nothing about most of their life histories and community relations.

Naturalists want to know about more than naming and classifying. Ecology, environmental science, conservation biology, and evolutionary biology engage the interests of naturalists. Fields as disparate as molecular genetics and wildlife management find common ground through their relationship to the naturalist tradition.

As literacy has increased and communication technology opened new possibilities, the audience for natural history has expanded. Natural history has always been a subject that has attracted a broad, reading public. Although the specialization of the nineteenth century distanced research scientists from the general reader, the link between them never entirely disappeared. All through the last century an ever-growing and appreciative audience consumed natural history books, articles, and illustrations. The great natural history museums constructed at the end of the nineteenth century had extensive public outreach programs that popularized their subject. Zoos and botanical gardens became favorite leisure-time sites and were—and remain—among the major attractions of urban centers. National parks, wildlife refuges, and wildlife parks served a similar function.

Advances in communication have created new opportunities for popularizing natural history; the most dramatic, of course, being the camera, motion pictures, and television. Starting with *Seal Island* in 1948, Walt Disney began producing documentaries on natural history subjects which became standard entertainment for the next two generations. As family entertainment, nature stories and documentaries have a well-established niche in the contemporary media and find extensive use in educational institutions. Cable television makes it possible to view nature programs twenty-four hours a day. Children can sit in their living rooms and watch nature shows featuring exotic species in their natural habitat which one hundred years ago could be observed only by the most adventurous travelers, often at great risk.

Popular natural history has not been completely taken over by television and film. Reflecting our increasing encroachment on the environment, much natural history writing in the second half of the twentieth century presents environmental concerns. Aldo Leopold proposed a land ethic in his widely appreciated *A Sand County Almanac* (1949), and Rachel Carson's *Silent Spring* (1962), which called attention to the environmental hazards of pesticides, showed how broad the reading public could be, even on such a serious biological subject.

Not all twentieth-century nature writing has centered on environmental issues. A rich tradition stretching back to Henry David Thoreau's *Walden* (1854)
ponders our place in nature and the individual's spiritual relationship to it. An-
nie Dillard's best-selling *Pilgrim at Tinker Creek* (1974) established her reputation as a leading contemporary nature writer in this genre. Highly sophisticated naturalists, such as Stephen Jay Gould, produce detailed and accurate
essays with a strong consciousness of the cultural dimension of their science.
Gould's articles weave history, nature, and society into stories that instruct and
entertain. And like E. O. Wilson, his colleague at Harvard, Gould appreciates
the importance of the naturalist tradition.

Natural history has been central to the life sciences for more than two centuries. Its importance continues. The need to document nature, the need to
understand its underlying regularities, and the need to construct an overall picture remain as important today as they ever were. Naturalists have a vast catalog to complete and a broad tableau to envision: one that includes and has considerable relevance for that uniquely reflective species, *Homo sapiens.*

Suggested Further Reading

For general treatment of the history of natural history see N. Jardine, J. A. Secord, and E. C. Spary, eds., *Cultures of Natural History* (Cambridge: Cambridge University Press, 1996), which has interesting articles on natural history from the Renaissance to the twentieth century. David Allen, *The Naturalist in Britain* (Princeton: Princeton University Press, 1994), deals with the British side of the history but contains a wealth of general information. John Foster, ed., *Nature in Ireland* (Dublin: Lilliput Press, 1997), focuses on Ireland but has essays of broader interest. For a thoughtful discussion of the modern sensibility toward nature see Keith Thomas, *Man and the Natural World: A History of the Modern Sensibility* (New York: Pantheon Books, 1983).

1 Collecting, Classifying, and Interpreting Nature

On Linnaeus see James Larson, *Reason and Experience: The Representation of Natural Order in the Work of Carl von Linné* (Berkeley: University of California Press, 1971), Tore Frängsmyr, ed., *Linnaeus: The Man and His Work* (Berkeley: University of California Press, 1983), Wilfrid Blunt, *The Compleat Naturalist: A Life of Linnaeus* (New York: Viking Press, 1971), Heinz Goerke, *Linnaeus* (New York: Scribner's, 1973), Londa Schiebinger, *Nature's Body: Gender in the Making of Modern Science* (Boston: Beacon Press, 1993), and Frans Stafleu, *Linnaeus and the Linnaeans: The Spreading of Their Ideas in Systematic Botany, 1735–1789* (Utrecht: A. Oosthoek's Uitgeversmaatschappij, 1971). For an interesting discussion of the Dutch background of natural history see Simon Schama, *The Embarrassment of Riches: An Interpretation of Dutch Culture in the Golden Age* (Berkeley: University of California Press, 1988). For a general treatment of eighteenth-century developments in the life sciences see James Larson, *Interpreting Nature: The Science of Living Form from Linnaeus to Kant* (Baltimore: Johns Hopkins University Press, 1994). For a general discussion of Buffon's life and work see Jacques Roger, *Buffon: A Life in Natural History* (Ithaca: Cornell University Press, 1997), and Otis Fellows and Stephen Milliken, *Buffon* (New York: Twayne Publishers, 1972). An excellent discussion of the history of natural history collections in the period before Buffon is in Paula Findlen, *Possessing Nature* (Berkeley: University of California Press, 1994).

2 New Specimens

For a description of the growth of natural history collections see Paul Lawrence Farber, *Discovering Birds: The Emergence of Ornithology as a Scientific Discipline, 1760–1850* (Baltimore: Johns Hopkins University Press, 1996), and David Allen, *The Naturalist in Britain* (Princeton: Princeton University Press, 1994). Scientific expeditions of the time are discussed in Frédéric Mauro, *L'Expansion Européenne (1600–1870)* (Paris: Presses Universitaires de France, 1967), John Dunmore, *French Explorers in the Pacific* (Oxford: Oxford University Press, 1965), Agnes Beriot, *Grand voliers autour du monde: Les voyages scientifiques 1760–1850* (Paris: Port Royal, 1962), and Daniel Baker, *Explorers and Discoverers of the World* (Detroit: Gale, 1993). Joseph Banks has been the object of many studies—see Hector Cameron, *Sir Joseph Banks* (Sydney: Angus and Robertson, 1966), Charles Lyte, *Sir Joseph Banks: Eighteenth-Century Explorer, Botanist, and Entrepreneur* (Newton Abbot: David and Charles, 1980), and Patrick O'Brian, *Joseph Banks: A Life* (London: C. Harvill, 1987). On William Swainson see Sheila Natusch and Geoffrey Swainson, *William Swainson of Fern Grove F.R.S., F.L.S., &c.: The Anatomy of a Nineteenth-Century Naturalist* (Wellington: Published by the authors with the aid of the New Zealand Founders Society, 1987). On Humboldt's influence see Susan Faye Cannon, *Science in Culture: The Early Victorian Period* (New York: Science History Publications, 1978), and articles by Malcolm Nicholson, who explores some of Humboldt's impact on naturalists who ultimately contribute to the emergence of ecology: "Alexander von Humboldt: Humboldtian Science and the Origin of the Study of Vegetation," *History of Science* 25(1987):167–94, and "Humboldtian Plant Geography after Humboldt: The Link to Ecology," *History of Science* 29(1996):289–310.

3 Comparing Structure

On Cuvier see William Coleman, *Georges Cuvier Zoologist: A Study in the History of Evolution Theory* (Cambridge: Harvard University Press, 1964), for his science, and for his career see Dorinda Outram, *Georges Cuvier: Vocation, Science and Authority in Post-Revolutionary France* (Manchester: Manchester University Press, 1984). On Geoffroy Saint-Hilaire see Théophile Cahn, *La vie et l'oeuvre d'Étienne Geoffroy Saint-Hilaire* (Paris: Presses Universitaires de France, 1962). For the Cuvier–Geoffroy Saint-Hilaire debate see Toby Appel, *The Cuvier-Geoffroy Debate* (Oxford: Oxford University Press, 1987). The classic discussion of the debates in comparative anatomy is E. S. Russell, *Form and Function* (London: John Murray, 1916). On fossils see Martin Rudwick, *The Meaning of Fossils* (New York: Neal Watson, 1976). Embryology and recapitulation are discussed in Stephen Jay Gould, *Ontology and Phylogeny* (Cambridge: Harvard University Press, 1977), and in Robert J. Richards, *The Meaning of Evolution* (Chicago: University of Chicago Press, 1992). For an interesting account of the history of natural classification in botany see Peter F. Stevens, *The Development of Biological Systematics: Antoine-Laurent de Jussieu, Nature, and the Natural System* (New York: Columbia University Press, 1994). Louis Agassiz's life is described by Edward Lurie in *Louis Agassiz: A Life in Science* (Chicago: University of Chicago Press, 1960), and in Mary P. Winsor, *Reading the Shape of Nature: Comparative Zoology at the Agassiz Museum* (Chicago: University of Chicago Press, 1991). For Lamarck see Richard Burkhardt Jr., *The Spirit of System* (Cambridge: Harvard University Press, 1995), and Pietro Corsi, *The Age of Lamarck: Evolutionary Theories in France 1790–1830* (Berkeley: University of California Press, 1988). Janet Browne, *The Secular Ark: Studies in the History of Biogeography* (New Haven: Yale University Press, 1983), discusses biogeography from Linnaeus to Darwin. On Owen see Nicolaas A. Rupke, *Richard Owen: Victorian Naturalist* (New Haven: Yale University Press, 1994).

4 New Tools and Standard Practices, 1840–1859

For Bonaparte see the detailed biography by Patricia Stroud, *The Emperor of Nature: Charles Lucien Bonaparte and His World* (Philadelphia: University of Pennsylvania Press, 2000). On technical innovations in natural history see Paul Lawrence Farber, *Discovering Birds: The Emergence of Ornithology as a Scientific Discipline, 1760–1850* (Baltimore: Johns Hopkins University Press, 1996), David Allen, *The Naturalist in Britain* (Princeton: Princeton University Press, 1994), and Karen Wonders, *Habitat Diorama: Illusions of Wilderness in Museums of Natural History*, Acta Universitatis Upsaliensis, Figura Nova Series 25 (Uppsala: University of Uppsala, 1993). On scientific illustration see Ann Blum, *Picturing Nature: American Nineteenth-Century Zoological Illustration* (Princeton: Princeton University Press, 1993), C. E. Jackson, *Bird Illustrators: Some Artists in Early Lithography* (London: Witherby, 1975), and David Knight, *Zoological Illustration: An Essay towards a History of Printed Zoological Pictures* (Folkstone, England: Dawson, 1977). Also see Isabella Tree, *The Ruling Passion of John Gould: A Biography of the Bird Man* (London: Barrie & Jenkins, 1991), and Alice Ford, *John James Audubon* (Norman: University of Oklahoma, 1964). For the role of women in natural history, see Barbara Gates, *Kindred Nature: Victorian and Edwardian Women Embrace the Living World* (Chicago: University of Chicago Press, 1998). The context of early American natural history is discussed in Charlotte Porter, *The Eagle's Nest: Natural History and American Ideas, 1812–1842* (Tuscaloosa: University of Alabama Press, 1986).

5 Darwin's Synthesis

The literature on Darwin is enormous. Janet Browne's *Charles Darwin: Voyaging* (New York: Alfred A. Knopf, 1995) is the first volume in an excellent biography that gives scientific, social, and personal dimensions of Darwin's life. Adrian Desmond and James Moore, *Darwin: The Life of a Tormented Evolutionist* (New York: Warner Books, 1991), gives a valuable description of Darwin's intellectual development in the context of British history. David Kohn, ed., *The Darwinian Heritage* (Princeton: Princeton University Press, 1985), is an introduction to the "Darwin Industry." Michael Ruse, *The Darwinian Revolution: Science Red in Tooth and Claw* (Chicago: University of Chicago Press, 1979), is a general introduction to the scientific issues involved in the Darwin story. Wallace is commented upon in many of the Darwin Industry works. A sympathetic biography of Wallace is Wilma George, *Biologist Philosopher: A Study of the Life and Writing of Alfred Russel Wallace* (New York: Abelard-Schuman, 1964). On Wallace also see John Langdon Brooks, *Just before the Origin: Alfred Russel Wallace's Theory of Evolution* (New York: Columbia University Press, 1984), and H. Lewis McKinney, *Wallace and Natural Selection* (New Haven: Yale University Press, 1972). On the reception of Darwin's ideas see David L. Hull, *Darwin and His Critics: The Reception of Darwin's Theory of Evolution by the Scientific Community* (Cambridge: Harvard University Press, 1973), Thomas F. Glick, ed., *The Comparative Reception of Darwinism* (Austin: University of Texas Press, 1974), and Peter J. Bowler, *The Eclipse of Darwinism: Anti-Darwinian Evolution Theories in the Decades around 1900* (Baltimore: Johns Hopkins University Press, 1983). The controversy over the age of Earth is examined in Joe Burchfield, *Lord Kelvin and the Age of the Earth* (New York: Science History Publications, 1975). An interesting discussion of the extension of evolutionary ideas is Peter J. Bowler, *Life's Splendid Drama: Evolutionary Biology and the Reconstruction of Life's Ancestry 1860–1940* (Chicago: University of Chicago Press, 1996).

6 Studying Function

On the emergence of physiology as a discipline see John E. Lesch, *Science and Medicine in France: The Emergence of Experimental Physiology, 1790–1855* (Cambridge: Harvard University Press, 1984), Frederic Lawrence Holmes, *Claude Bernard and Animal Chemistry* (Cambridge: Harvard University Press, 1974), and Joseph Schiller, *Physiology and Classification* (Paris: Maloine, 1980). For the professors in the nineteenth century see *Centenaire de la fondation du Muséum d'Histoire Naturelle: Volume Commémoratif* (Paris: Imprimerie Nationale, 1893), iii–vii. An interesting discussion of the attitudes toward experimentation on animals is in Richard French, *Antivivisection and Medical Science in Victorian Society* (Princeton: Princeton University Press, 1975). For background on the competition over funds in the life sciences see Robert Fox, "Science, the University, and the State in Nineteenth-Century France," in Gerald L. Geison, ed., *Professions and the French State, 1700–1900* (Philadelphia: University of Pennsylvania Press, 1984), 66–145, Robert Fox, ed., *The Organization of Science and Technology in France 1808–1914* (Cambridge: Cambridge University Press, 1980), and Harry Paul, *From Knowledge to Power: The Rise of the Science Empire in France, 1860–1939* (Cambridge: Cambridge University Press, 1985). On the later development of experimental biology see William Coleman, *Biology in the Nineteenth Century: Problems of Form, Function, and Transformation* (New York: Wiley, 1971), Garland Allen, *Life Science in the Twentieth Century* (New York: Wiley, 1975), Jane Maienschein, *Transforming Traditions in American Biology, 1880–1915* (Baltimore: Johns Hopkins University Press, 1991), Gerald L. Geison, ed., *Physiology in the American Context, 1850–1940* (Bethesda, Md.: American Physiological Society, 1987).

7 Victorian Fascination

There are many good sources on the great museums, zoos, and botanical gardens. Richard D. Altick, *The Shows of London* (Cambridge: Harvard University Press, 1978), Harriet Ritvo, *The Animal Estate* (Cambridge: Harvard University Press, 1987), and Ritvo's *The Platypus and the Mermaid and Other Figments of the Classifying Imagination* (Cambridge: Harvard University Press, 1997) provide a background on popular interest in animals and animal displays. Useful discussions can be found in Gustave Loisel, *Histoire des ménageries de l'antiquité à nos jours*, 3 vols. (Paris: Doin, 1912), Susan Sheets-Pyenson, *Cathedrals of Science: The Development of Colonial Natural History Museums during the Late Nineteenth Century* (Montreal: McGill-Queens University Press, 1988), Charles Coleman Sellers, *Mr. Peale's Museum: Charles William Peale and the First Popular Museum of Natural Science and Art* (New York: W. W. Norton, 1980), Joel L. Orosz, *Curators and Culture: The Museum Movement in America, 1740–1870* (Tuscaloosa: University of Alabama Press, 1990), Lucile H. Brockway, *Science and Colonial Expansion: The Role of the British Royal Botanic Gardens* (New York: Academic Press, 1979), Wilfred Blunt, *In for a Penny: A Prospect of Kew Gardens, Their Flora, Fauna and Falballas* (London: Hamish

Hamilton, 1978), and A. H. Saxon, *P. T. Barnum: The Legend and the Man* (New York: Columbia University Press, 1989).

For a discussion of the Jardin zoologique d'acclimatation see Michael Osborne, *Nature, the Exotic, and the Science of French Colonialism* (Bloomington: Indiana University Press, 1994). An engaging history of the Bronx Zoo can be found in William Bridges, *Gathering of Animals: An Unconventional History of the New York Zoological Society* (New York: Harper & Row, 1974). Amateur botany is well described in Elizabeth B. Keeney, *The Botanizers: Amateur Scientists in Nineteenth-Century America* (Chapel Hill: University of North Carolina Press, 1992). Interest in entomological collecting is covered by Willis Conner Sorensen, *Brethren of the Net: American Entomology, 1840–1880* (Tuscaloosa: University of Alabama, 1995). The background of the British Museum (Natural History) is discussed in Nicolaas A. Rupke, *Richard Owen: Victorian Naturalist* (New Haven: Yale University Press, 1994), and considerable information is in Albert E. Gunther, *A Century of Zoology at the British Museum Through the Lives of Two Keepers, 1815–1914* (London: Dawsons of Pall Mall, 1975). The American Museum of Natural History is discussed in Douglas J. Preston, *Dinosaurs in the Attic: An Excursion into the American Museum of Natural History* (New York: St. Martin's Press, 1986), and Ronald Rainger, *An Agenda for Antiquity: Henry Fairfield Osborn and Vertebrate Paleontology at the American Museum of Natural History, 1890–1935* (Tuscaloosa: University of Alabama Press, 1991). On Huxley see Adrian Desmond, *Huxley: From Devil's Disciple to Evolution's High Priest* (Reading, Pa.: Addison-Wesley, 1997). For articles that discuss the emergence of biology as an academic subject see Ronald Rainger, Keith Benson, and Jane Maienschein, eds., *The American Development of Biology* (Philadelphia: University of Pennsylvania Press, 1988).

An interesting examination of ornithology in the years after Audubon is Mark Barrow, *A Passion for Birds: American Ornithology after Audubon* (Princeton: Princeton University Press, 1998). The feather trade is covered in Robin W. Doughty, *Feather Fashions and Bird Preservation: A Study in Nature Protection* (Berkeley: University of California Press, 1975). An enormous literature exists on the history of nature writing. Of interest on the late nineteenth century and turn of the century are Lynn L. Merrill, *The Romance of Victorian Natural History* (Oxford: Oxford University Press, 1989), and Ralph H. Lutts, *The Nature Fakers: Wildlife, Science, and Sentiment* (Golden, Colo.: Fulcrum Publishing, 1990).

8 New Synthesis

On experimental taxonomy see Joel Hagen, "Experimentalists and Naturalists in Twentieth-Century Botany: Experimental Taxonomy, 1920–1950," *Journal of the History of Biology* 17, no. 2 (1984): 249–70. Julian Huxley's many contributions to modern biology are discussed in C. Kenneth Waters and Albert Van Helden, eds., *Julian Huxley: Biologist and Statesman of Science* (Houston: Rice University Press, 1992). On the Modern Synthesis see Ernst Mayr and William B. Provine, eds., *The Evolutionary Synthesis: Perspectives on the Unification of Biology* (Cambridge: Harvard University Press, 1980), Mark Adams, ed., *The Evolution of Theodosius Dobzhansky* (Princeton: Princeton University Press, 1994), Vassiliki Betty Smocovitis, *Unifying Biology: The Evolutionary Synthesis and Evolutionary Biology* (Princeton: Princeton University Press, 1996), and Joseph Cain, "Common Problems and Cooperative Solutions: Organizational Activity in Evolutionary Studies, 1936–1947," *ISIS* 84, no. 1 (1993): 1–25. On Ernst Mayr see Walter J. Bock, "Ernst Mayr, Naturalist: His Contributions to Systematics and Evolution," *Biology and Philosophy* 9, no. 3 (1994): 267–327, and Joseph Cain, "Ernst Mayr as Community Architect: Launching the Society for the Study of Evolution and the Journal *Evolution*," *Biology and Philosophy* 9, no. 3 (1994): 387–427.

9 The Naturalist as Generalist

Wilson's views on natural history are developed in his autobiography, Edward O. Wilson, *Naturalist* (Washington, D.C.: Island Press, 1994), and in his "The Coming Pluralization of Biology and the Stewardship of Systematics," *BioScience* 39(1989): 242–45. Wilson's broader views on natural history are discussed in *Biophilia* (Cambridge: Harvard University Press, 1984), and *Consilience: The Unity of Knowledge* (New York: Alfred A. Knopf, 1998). For the "new natural history" see the afterword in Sharon Kingsland, *Modeling Nature: Episodes in the History of Population Biology,* 2nd ed. (Chicago: University of Chicago Press, 1995). Conservation biology is discussed in Peter Davis, *Museums and the Natural Environment: The Role of Natural History Museums in Biological Conservation* (London: Leicester University Press, 1996). An interesting study of nature writing can be found in Peter Fritzell, *Nature Writing and America: Essays upon a Cultural Type* (Ames: Iowa State University Press, 1990).

Index

Library of Congress-in-Publication Data

Farber, Paul Lawrence, 1944–
    Finding order in nature : the naturalist tradition from Linnaeus to E. O.
Wilson / Paul Lawrence Farber.
        p. cm. — (Johns Hopkins introductory studies in the history of science)
    Includes bibliographical references (p.) and index.
    ISBN 0-8018-6389-9 — ISBN 0-8018-6390-2 (pbk. : alk. paper)
    1. Natural history—History.   I. Title.   II. Series.

    QH15 .F27 2000
    508'.09—dc21   99-089621